T0270995

Cambridge Studies in Biotechnology

Editors: Sir James Baddiley, N. H. Carey, I. J. Higgins,
W. G. Potter

9 Biotechnology of microbial exopolysaccharides

9 Biotechnology of microbial exopolysaccharides

Other titles in this series

Biotechnology of microbial exopolysaccharides

IAN W. SUTHERLAND

Department of Microbiology, Edinburgh University

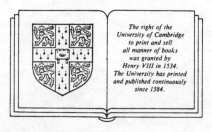

The right of the
University of Cambridge
to print and sell
all manner of books
was granted by
Henry VIII in 1534.
The University has printed
and published continuously
since 1584.

CAMBRIDGE UNIVERSITY PRESS

Cambridge

New York   Port Chester

Melbourne   Sydney

CAMBRIDGE UNIVERSITY PRESS
Cambridge, New York, Melbourne, Madrid, Cape Town, Singapore, São Paulo

Cambridge University Press
The Edinburgh Building, Cambridge CB2 8RU, UK

Published in the United States of America by Cambridge University Press, New York

www.cambridge.org
Information on this title: www.cambridge.org/9780521363501

© Cambridge University Press 1990

First published 1990
This digitally printed version 2008

A catalogue record for this publication is available from the British Library

ISBN 978-0-521-36350-1 hardback
ISBN 978-0-521-06394-4 paperback

Contents

Preface

Industrial interest in microbial polysaccharides has been stimulated by their unique properties and the opportunity to provide a guaranteed supply of material of constant quality and stable price. One must set against such positive aspects the relatively high costs of the product, of process development, and of downstream processing, especially if the intention is to provide material with approval for food usage.

The aim of this book is to present information relating to microbial exopolysaccharides which have actual or potential industrial or medical importance, rather than to provide comprehensive coverage of the whole field. It indicates the mechanisms by which these polymers are synthesised, as well as techniques used in their chemical and physical characterisation.

There has been a marked upsurge of interest in microbial exopolysaccharides in recent years, from biologists and non-biologists alike. In particular, recent studies on the physicochemical properties of polysaccharide solutions are providing a new insight into the physical structures of these polymers and furnishing the industrialist with a clearer indication of their useful properties. The increased interest in microbial polysaccharides mirrors a growth in the use of water-soluble polymers generally, and also an appreciation of the environmental advantages to be gained from use of water-soluble rather than solvent-based systems.

As many students now receive little instruction in the chemistry of carbohydrates and related molecules, readers may need to refer to a suitable text on that subject. It should, however, be remembered that in many respects the physical properties of polysaccharides are frequently not greatly dissimilar from those of DNA and RNA. The molecular biologist using enzymes to sequence nucleic acids and proteins may even find that a very similar approach has been used for many years in structural studies of polysaccharides. It is hoped that this volume will provide background information for the biologist, chemist, industrialist and pharmacologist, who may wish to use microbial polysaccharides and requires a suitable introduction to the subject. It is intended for those who have some training in biotechnology or one of its enabling disciplines, but who do not have a specialised knowledge of microbial polysaccharide structure and synthesis.

As the structure, production and synthesis of exopolysaccharides have been extensively reviewed in recent years, the reader will not find detailed references provided for much of this material. Instead, he or she will find a list of relevant review material for further reading at the end of each chapter. Most of the references that are included cover material of recent date (mainly 1986 onwards) which has not yet reached the review literature or which has come from less commonly searched areas. Such references are provided in full in the bibliography following the final chapter.

I am indebted to many friends and colleagues in the polysaccharide world for their helpful advice and comments and for the gift of polymers. Their willingness to permit reproduction of figures and to provide photographic material is gratefully acknowledged. I am also grateful to Karen for her critical reading of the manuscript and correction of grammatical and typographical errors, and to Ann and Karen for their tolerance during the gestation period of this volume.

I.W.S.

1 Introduction and definition

1.1. Introduction

The surface of the microbial cell is a rich source of carbohydrate-containing molecules. Some of these are unique types, confined to a limited range of microorganisms. These are the components of the microbial cell walls such as yeast mannans, bacterial teichoic and teichuronic acids, lipopolysaccharides and peptidoglycan. However, in addition to these wall components, polysaccharides may be found either associated with other surface macromolecules or totally dissociated from the microbial cell. These are **exopolysaccharides**, extracellular polysaccharides showing considerable diversity in their composition and structure. Some of these polymers may bear a strong chemical similarity to cell-wall components, but the majority are distinct chemical structures totally unrelated to cellular constituents.

Exopolysaccharides occur widely, especially among prokaryotic species, both among those that are free-living saprophytes and among those that are pathogenic to humans, animals and plants. Most microalgae yield some type of exopolysaccharide but they are less common among yeasts and fungi. However, some of those isolated from fungi do possess interesting physical and pharmacological properties.

Definition of exopolysaccharides is more difficult than definition of the carbohydrate-containing polymers found in microbial walls. The term exopolysaccharide has been widely used to describe polysaccharides found external to the structural outer surface of the microbial cell and it can be applied to polymers of very diverse composition and of different physical types. The term *glycocalyx*, introduced by Costerton, fails to differentiate between the different chemical entities found at the microbial surface. It has been used to represent a complex array of macromolecular species inlcuding components which are truly extracellular, together with wall polysaccharides and many other non-carbohydrate-containing chemical species. Although it may describe structures seen under the microscope or electron microscope, it is inadequate in chemical terms.

The exopolysaccharides do not in themselves normally contribute to microbial structure; the other components of the cell surface are unaltered if exopolysaccharides are absent. They do, however, form structures which can be recognised *per se* by either light or electron microscopy. In a few microbial genera, they form components of more complex structures which may be involved in different morphogenetic cycles such as those found in

1

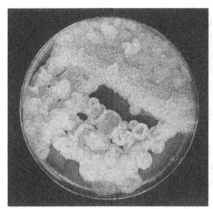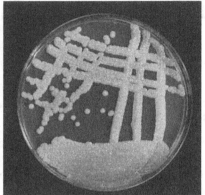

Fig. 1.1. The exotic colonial appearance of two polysaccharide-producing pseudomonads.

the families Azotobacteriaceae and the Myxobacteriaceae. In each of these bacterial groups, exopolysaccharides are associated with normal vegetative cells and with resting cells in the form of microcysts. In the Azotobacteriaceae, microcysts and vegetative cells differ little in shape and both are surrounded by exopolysaccharide of similar composition. In the Myxobacteriaceae, the microcysts and polysaccharide are contained within complex fruiting bodies; the vegetative cells also excrete considerable quantities of extracellular polysaccharide and resemble the Myxomycetes in leaving 'slime trails' on the surface of solid media.

1.2.   Definition

The presence of exopolysaccharides associated with microbial cells grown on solid surfaces is frequently recognisable from the mucoid colony morphology. This may be linked, in some bacterial isolates, to unusual appearance (Fig. 1.1). In liquid medium, exopolysaccharide-producing cultures may become very viscous or, exceptionally, may solidify as a gel. The exopolysaccharide may form part of a *capsule* firmly attached to the bacterial cell surface. Alternatively, it may be observed as loose *slime* secreted by the microorganisms but not directly attached to the cell. On solid surfaces exposed to aqueous environments, whether within the human or animal body, in fresh water or in the oceans, bacterial growth is seen as *biofilms*, in which the microbial cells are associated with large amounts of exopolysaccharide. Unfortunately, many of the descriptive reports of extracellular polysaccharides in the laboratory and in natural environments have failed to recognise the relationship between the physical forms of these macromolecules and the physiological conditions present. Changes in the growth conditions can drastically alter the composition, physical properties and organisation of the polysaccharides at the bacterial surface.

Visualisation of exopolysaccharides under the light microscope is possible either through negative staining or through the use of specialised stains, which frequently utilise the polyanionic function of many of these polymers. Negative staining has the advantage of distinguishing between capsules and extracellular slime. Techniques such as scanning electron microscopy (SEM) frequently provide strong evidence for the presence of exopolysaccharide, but are not capable of providing confirmatory chemical evidence. Equally, the modern techniques of transmission electron microscopy (TEM) can reveal considerable information about the surface structures, including the polysaccharides involved, but are unable to distinguish between different chemotypes; neither do they have sufficient resolution to provide much information on the microstructure of the bacterial capsule. Unfortunately, the very high water content of the polysaccharides, in excess of 99%, makes preparation and resolution difficult. However, improved electron microscopic preparation methods, including pretreatment with anticapsular IgG and freeze-etching, can reveal the exopolysaccharide in an uncollapsed state. Capsules can be visualised as being fibrous in composition. Thin sections of gelatine-enrobed bacteria show fibrous strands extending radially from the bacterial surface (Fig. 1.2). Apparent association between the fibres may represent the ordered form of the polymer molecules. Improved electron microscopic methods can also be used to study the structure of pure polysaccharide when prepared in aqueous solution. This *may* provide supporting evidence for their conformation in solution and for intramolecular interaction.

1.3.   Composition

Exopolysaccharides are primarily composed of carbohydrates, but in addition to the various sugars, there may be organic and inorganic substituents. The carbohydrates found in microbial exopolysaccharides are extremely diverse. Most of the sugars are those commonly found in animal and plant polysaccharides. D-glucose, D-galactose and D-mannose in the pyranose forms are present in many exopolysaccharides. The 6-deoxyhexoses, L-fucose and L-rhamnose, are also frequently present. A distinction between eukaryotes and prokaryotes can be seen in the presence of pentoses. Eukaryotic polysaccharides may contain pentoses such as D-ribose or D-xylose, but they are of less common occurrence in extracellular polymers derived from prokaryotes. The Cyanobacteria provide an exception. In this group of bacteria, pentoses may be found in the sheath polysaccharides.

In addition to the more common monosaccharides, some polysaccharides may contain one or more rare sugars. These may include L-hexoses or furanose forms of the hexoses, glucose and galactose. There are also

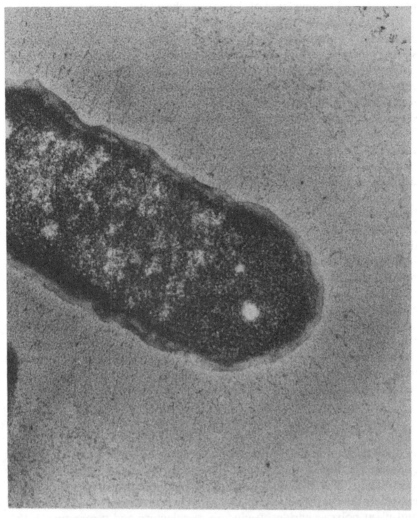

Fig. 1.2. Electron micrograph of a capsulate *Escherichia coli* strain K29. The bacteria are fixed with glutaraldehyde, enrobed with gelatine and embedded. The capsular material forms a layer about 600 nm thick around the bacterium. (Reproduced from Bayer *et al.* (1985) with permission.)

various *N*-acetylamino sugars, although as yet these have not been found in any of the microbial polysaccharides of industrial importance. Amino sugars are present in polysaccharides from species such as *Escherichia coli* but absent from the products of genera such as *Xanthomonas* or *Klebsiella*. The commonest amino sugars are *N*-acetyl-D-glucosamine and *N*-acetyl-D-galactosamine. *N*-acetyl-D-mannosamine is occasionally found and there are also several rarer amino sugars such as fucosamine and talosamine.

Inevitably, as a wider range of microbial sources of exopolysaccharides is examined, new monosaccharides will be discovered.

Many or perhaps even most microbial exopolysaccharides are polyanionic in nature. This results from the presence in many of them of uronic acids, with D-glucuronic acid being easily the commonest. D-galacturonic acid is less common and only a small number of polysaccharides containing D-mannuronic acid have been identified. Although most polysaccharides contain only a single uronic acid, D-mannuronic acid is usually present in bacterial alginates together with L-guluronic acid. Some polysaccharides from *Rhizobium* species are known to contain both D-glucuronic acid and D-galacturonic acid. An exopolysaccharide with L-iduronic acid has recently been reported from the anaerobic rumen bacterium *Butyrivibrio fibrisolvens*. This sugar is found in the sulphated proteoglycans present in higher eukaryotes but had not until now been reported in prokaryotic material. Hexosaminuronic acids have been identified in a wide range of bacterial polysaccharides. Indeed, the exopolysaccharide from *Staphylococcus aureus* strain M contains both 2-acetamido-2-deoxy-D-fucose and 2-acetamido-2-deoxy-D-galacturonic acid.

Another unusual sugar found in exopolysaccharides is ketodeoxyoctonic acid (KDO), 3-keto-deoxy-D-*manno*octulosonic acid. This sugar is of almost universal occurrence in the core portion of the lipopolysaccharides from the walls of Gram-negative bacteria. It was originally thought to be limited to these polymers but has recently been identified in eukaryotic material and as a component of several bacterial exopolysaccharides from strains of *Escherichia coli*.

1.4.   Organic substituents

In addition to carbohydrates, the microbial exopolysaccharides contain various ester-linked substituents and pyruvate ketals. These are very widely found in bacterial exopolysaccharides, although they have as yet seldom been reported from eukaryotic polymers. Acetate, as an ester-linked substituent, does not contribute to the overall charge on the polysaccharide molecule, whereas the pyruvate ketals add to the anionic nature of the polymers. Pyruvate is normally present in stoichiometric ratios with the sugar components of the exopolysaccharide and is usually attached to a neutral hexose. It is, however, occasionally attached to uronic acid, and one exopolysaccharide from *Klebsiella aerogenes* contains a pyruvylated methylpentose. The pyruvate is commonly present along with uronic acid residues, both contributing to the overall anionic charge. Indeed, in an exopolysaccharide secreted by *Rhizobium leguminosarum*, *R. trifolii* and *R. phaseoli*, the octasaccharide repeat unit is unusual as it contains two moles of D-glucuronic acid, one mole of pyruvate linked as a ketal to D-galactose, and another mole similarly linked to D-glucose (Zevenhuizen, 1984). The two pyruvylated hexoses are the terminal and subterminal residues, respectively, of a tetrasaccharide side-chain. In some exopolysaccharides, pyruvate alone is responsible for the anionic nature of the polymer. This is true of a polysaccharide from *E. coli* (Anderson *et al.*, 1987) and of material from *Zoogloea ramigera*. Both of thesê polymers contain D-glucose and D-galactose as the only monosaccharides, the pyruvate being linked to galactose. Another similar polymer, obtained from a number of isolates of *Pseudomonas marginalis*, contains the same sugars along with pyruvate and succinate (Osman and Fett, 1989). The commonest ester-linked component is acetate, but propionyl groups have also been found in *E. coli* K14 exopolysaccharide and ester-linked glyceryl residues form part of the structure of the polymer from *Pseudomonas elodea* before it is converted to the deacylated form which is sold as gellan or gelrite. In each case, these unusual esters are present as well as acetate and will contribute to the lipophilic nature of part of the polysaccharide molecule. Succinyl groups, as the half ester, are also found along with acetyl groups and may represent up to 7% of the exopolysaccharide weight from certain *Rhizobium* species and from *Agrobacterium* species. Also present in some bacterial polysaccharides, as ester-linked residues attached to D-galactose, is 3-hydroxybutanoate. This substituent is present in material from *R. trifolii* and *R. leguminosarum*. Lactyl groups are found in some *Klebsiella aerogenes* exopolysaccharides. Although formate was at one time thought to be another acyl group present in some *K. aerogenes* polysaccharides, its presence in this group of exopolysaccharides has now been discounted after re-examination with fast atom bombardment spectroscopy.

The acyl content of some polysaccharides can be relatively high and, as well as increasing the lipophilicity of the molecule, may affect the capacity to interact with other polysaccharides and with cations.

In many exopolysaccharides the acyl groups are found in stoichiometric ratios relative to the monosaccharides present, but this is not always the case. Several polymers contain acyl substituents on alternate repeat units. Other polymers may be more random in their acylation and it should be remembered that most commercial preparations of xanthan only contain pyruvate on about 30% of the terminal side-chain mannosyl residues. In bacterial alginates, which lack any regular repeat structure, the acetyl content may be up to 15–20%, only the D-mannuronic acid residues being acetylated. Bacterial alginates also belong to the small group of polymers in which multiple acetylation of single sugar residues has been detected.

The presence of several amino acids in bacterial exopolysaccharides has recently been reported. Serine is found in the exopolysaccharide from *E. coli* K40; L-glutamic acid has been detected during a re-examination of *Klebsiella aerogenes* type 82 polysaccharide. Taurine is found as an ester-linked substituent in some other polysaccharides.

1.5.   Inorganic substituents

It was long thought that sulphate was limited to eukaryotic polysaccharides and proteoglycans, but it now appears that sulphate is present in some prokaryotic polymers. As yet, these are confined to species of Cyanobacteria. The extracellular sheath or flocculant produced by the cyanobacterial *Phormidium* species is one such sulphated heteropolysaccharide.

Phosphate is of much more widespread occurrence and is frequently found in bacterial polysaccharides of immunological significance. Many of the phosphorylated exopolysaccharides resemble the teichoic acids present in the walls of Gram-positive bacteria; indeed, many strains produce phosphate-containing wall polymers and extracellular polysaccharides. Although phosphate is absent from many of the Gram-negative bacterial polysaccharides investigated so far, including those from genera such as *Klebsiella*, *Rhizobium*, *Xanthomonas* and *Pseudomonas*, it has been identified in the products of a number of *E. coli* strains. As well as the inorganic components which form part of the structure of exopolysaccharides, it has to be remembered that all polyanionic polymers will normally be obtained in the salt form. These will represent a mixture of cations, although some may be bound more firmly than others. Thus some alginates bind the divalent cations calcium, barium and strontium very strongly, whereas polysaccharide XM6, from a strain of *Enterobacter*, favours the monovalent and divalent cations with ionic radius of about 1.0, i.e. Na^+ and Ca^{2+}. The ions associate with the polymer during its production but can be displaced by appropriate procedures (ion exchange, electrodialysis, etc.) to enable conversion into a free acid or uniform salt form. This is important in determining the physical properties of some of the exopolysaccharides but is not normally needed for commercial products. The range of non-carbohydrate substituents found in microbial exopolysaccharides is indicated in Table 1.1.

1.6.   Homopolysaccharides

Many microbial exopolysaccharides, including several of potential industrial importance, are homopolymers. These include a number of glucans, which, because of their different structures, possess significantly different properties even though the sole monosaccharide component is D-glucose. The fungal product scleroglucan has high viscosity whereas curdlan, a product of several bacterial species, is gel-forming, and bacterial cellulose is microcrystalline and insoluble. Curdlan and cellulose are composed of a single linkage type, but several other homopolysaccharides, including scleroglucan and pullulan, possess two types of glucosyl linkage. Dextran, also a glucan, is more complex and contains three different types of linkage

Table 1.1. *Non-carbohydrate substituents in exopolysaccharides*

Substituent	Linkage	EPS-producing bacterium
Organic acids		
Acetate	ester	very common
Glycerate	ester	*Pseudomonas elodea*
Hydroxybutanoate	ester	*Rhizobium trifolii, R. leguminosarum*
Propionate	ester	*Escherichia coli*
Pyruvate	ketal	very common
Succinate	ester	*Rhizobium* spp, *Agrobacterium* spp
Amino acids		
L-glutamate		*Klebsiella aerogenes* K82
Serine		*E. coli* K40
Inorganic acids		
Phosphate		common
Sulphate		cyanobacteria

(Table 1.2). Although most homopolysaccharides are composed of neutral sugars, a small number are polyanionic. Alginates composed solely of D-mannuronic acid can be obtained, and some bacterial homopolysaccharides are polysialic acids (Chapter 11).

1.7.   Heteropolysaccharides

The majority of microbial polysaccharides are probably heteropolysaccharides. These range from polymers with two sugar components to others with four or five monosaccharides. The possible range of structures and of resultant differences in properties is very great indeed because of the number of possible linkages and configurations. Each hexose can be α- or β-linked; in the pyranose or furanose form; and linked through the 2, 3, 4 or 6 position. However most of the polysaccharides that have been discovered so far are formed from two or three sugars and various acyl substituents. It is also probable that most of the polymers with industrial potential will contain certain types of linkage, at least in the main chain, as these will confer the desired physical properties (Chapter 8). However, a much wider range of structures can be expected among polysaccharides of medical interest.

1.8.   Visualisation of polysaccharides

Several methods of examining exopolysaccharides take advantage of the polyanionic nature of many of them. One such method involves cytochemical treatment with cationic ferritin. The complexes formed with the acid

Table 1.2. *Properties of dextrans*

Group	Linkage class	Intrinsic viscosity at 25 °C	Nature of polymer	Appearance of 1–2% aqueous solution
1	A,B	1.2–0.6	very cohesive, tough gum or flocculant	very turbid
2	A,C	0.5–0.2	fine or flocculant precipitate	opalescent solution
3a	A,B	0.9–0.5	fine or flocculant precipitate or dense gum	very turbid
3b	C	1.4–0.5	flocculant precipitate or dense gum	very turbid
4a	A,B	1.3–1.0	soft gum	slightly opalescent
4b	A,B	2.0–1.6	cohesive, stringy gum	slightly opalescent
4c	A,C	1.4–0.4	stringy or fluid gums	clear or slightly turbid
5a	A,B	1.2–0.6	short or stringy gums	turbid or slightly opalescent
5b	A,B	1.0–0.9	flocculant precipitate or short gum	slightly to very turbid

Fig. 1.3. Electron micrograph of a xanthan preparation showing the apparently double-stranded chain separating into two thinner strands. Prepared in 50% glycerol, 2 mM ammonium acetate, pH 7.0. Bar, 200 nm. (From Stokke *et al.* (1986) with permission.)

polysaccharides of *Klebsiella* species can then be readily seen. If the preparations are first interacted with homologous anticapsular serum and fixed with glutaraldehyde, the polymeric material is cross-linked and the polycationic ferritin cannot penetrate the complex but is seen on the capsular surface.

Recent studies have shown that xanthan, vacuum dried from a glycerol-containing solution, can be rotary shadowed and examined by electron microscopy. The polysaccharide can be seen as single- and double-stranded chains, together with partly dissociated double-stranded structures (Fig. 1.3). From the electron microscopic data and intrinsic viscosity measurements, the average relative molecular mass can be calculated.

Further reading

Sutherland, I. W. (1988). *Bacterial surface polysaccharides: structure and function. International Review of Cytology* **113**, 187–231.

2   Polysaccharide analysis and structural determination

2.1.   Introduction

The information that needs to be obtained in respect of defining microbial exopolysaccharides can be summarised under the headings of polymer composition and of structural analysis. Composition covers not only the monosaccharide components but also the various possible acyl groups and inorganic substituents. Structural analysis involves various approaches. These range from partial fragmentation by acid or enzymic hydrolysis to produce oligosaccharides, the structures of which must then themselves be determined, to methylation and sequence analysis.

2.2.   Analysis of polysaccharide composition

Exopolysaccharides are composed of three distinct types of monomer. They are of course predominantly carbohydrate in nature but, in addition to the various sugars, there may be organic and inorganic substituents. The individual sugars can seldom be analysed in the intact polymer although it is often possible to use colorimetric assays to quantitate specific types of sugars such as pentoses, 6-deoxyhexoses, heptoses, uronic acids and amino sugars. If the specific monosaccharides are to be determined quantitatively, they must first be released by hydrolysis.

Determination of the composition of microbial exopolysaccharides is essentially similar to that of any other comparable polymer. The polymer is hydrolysed with acid to yield the component monosaccharides, which are then identified and quantified. During hydrolysis, labile groups such as ester-linked components and ketals are likely to be removed and must be separately recovered and identified. A careful choice of hydrolysis conditions must also be made. Monosaccharides differ in their stability to acid at high temperatures; the glycosidic bonds vary considerably in their resistance to hydrolysis. The bonds between uronic acids or amino sugars and neutral hexoses are particularly stable. Although paper or thin-layer chromatography of underivatised monomers has been used effectively, high-performance liquid chromatography (HPLC) now provides a range of extremely rapid methods for separation and identification. Different columns are needed for different groups of monosaccharides if effective separation and quantification is to be obtained. Thus hexoses and 6-deoxyhexoses can be effectively separated on amino or cyanoamino

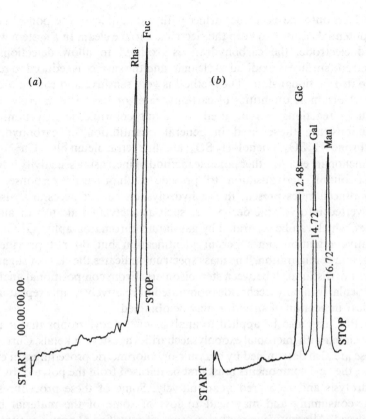

Fig. 2.1. The chromatographic separation of neutral sugars by HPLC on a SCX column in the Pb^{2+} form, using water as eluant: (*a*) 6-deoxyhexoses and (*b*) hexoses.

columns by using reversed-phase chromatography with acetonitrile–water mixtures as eluants. The different monosaccharides within each group are then further separated from each other on SCX (strong cation exchange) materials loaded with heavy metal ions such as lead, silver or calcium. Water is used as the mobile phase. The calcium form of SCX adsorbents may even permit the separation of the α- and β-anomers of the monosaccharides, a factor that has to be taken into consideration when quantifying the sugars. Uronic acids can be separated on SAX (strong anion exchange) columns. Detection of the monosaccharides is commonly through refractive index monitoring, but the sensitivity of the technique is relatively low and high sample loading of the columns is necessary. Correlation also has to be made for the response of the detector to different sugars (Fig. 2.1). A new method of detection is **pulsed amperometric detection (PAD)** in which the carbohydrates are oxidised by application of a voltage across an electrochemical cell. To avoid adsorption of the oxidation products onto the electrode surface with resultant loss of response, a cycle of potentials is used to keep the electrode surface clean. In a system with a gold electrode, the carbohydrate is oxidised to allow detection, the electrode surface is oxidised to clean it, and the surface is reduced to return it to its functional state. The method is very sensitive and can be used to detect picamole quantities of carbohydrates or less. Alternatively, post-column reactions can be used to permit colorimetric detection. The reactions are those used in general quantitation of carbohydrates: anthrone–H_2SO_4, phenol–H_2SO_4, alkaline tetrazoleum blue, Cu^{2+}-2,2'-bicinchonate etc. Another possible method of increasing sensitivity is to use post-column derivatisation to provide a fluorimetric response. The monosaccharides present in the hydrolysates can, if necessary, also be converted into volatile derivatives such as acetylated alditols or aldoni-triles, which can be separated by gas–liquid chromatography (GLC). The relative retention times permit identification but do not provide the absolute configuration. The mass spectrum indicates the class of sugar but does not distinquish between stereoisomers. From compositional analysis, particularly in polysaccharides composed of relatively simple repeat units, a clear indication of structure may be obtained.

HPLC can also be applied to analysis of the acyl groups that may be associated with microbial exopolysaccharides as esters or ketals. Currently, these are also determined by enzymic or colorimetric procedures. In either case, the acyl components must first be released from the polymer by mild hydrolysis and recovered quantitatively. Some of these procedures are time-consuming and may lead to loss of some of the material being analysed. Alternatively, methyl protons of esterified O-acetyl groups and pyruvate ketals can be detected by the use of ^{1}H NMR (nuclear magnetic resonance) spectroscopy without the necessity for hydrolysing the polysac-charides. The technique provides an accurate method for quantifying these substituents.

Various specific enzyme assays are available to determine free sugars after hydrolysis. Thus D-glucose may be determined through the use of D-glucose oxidase or through a combined use of hexokinase and glucose 6-phosphate dehydrogenase. D-galactose can be measured by using D-galactose oxidase. This enzyme can also be used for quantification of N-acetyl-D-galactosamine and D-galactosamine. Several other enzymic techniques are available and have the advantage of distinguishing between D-and L-forms of the monosaccharide. Pyruvate released hydrolytically from sugar ketals can also be determined enzymically by using the enzyme lactate dehydrogenase.

2.3.   Structural determination

The structural determination of polysaccharides is rather more compli-cated than that of proteins and nucleic acids. For each monosaccharide, one must determine the ring configuration (furanose or pyranose), the position of the linkage, the anomeric configuration (α or β) and the actual sequence of the monomers.

The classical method for obtaining much of this information is through methylation followed by hydrolysis, identification and quantification of the partially methylated sugars. All free hydroxy groups in the polysaccharide are methylated. This yields information on the ring size and on the substitution of the individual monosaccharides. The partly methylated sugars are also quantified; from the molar ratios, it should then be possible to derive the size and structure of the repeating unit. One major drawback found with earlier methods of methylation was the failure of the techniques then available to ensure total methylation. This led to the belief that many microbial exopolysaccharides were highly branched structures. Improved methods developed by Hakamori have overcome some but not all of the problems associated with polysaccharide methylation. The method allows complete methylation in a single step. Ideally, any uronic acids should first be reduced to neutral sugars before methylation. This also overcomes the problem of the highly stable aldobiouronic acid linkages, which are much more resistant to acid hydrolysis than are most other glycosidic bonds. Carboxyl reduction with borodeuteride permits distinction of the dideuterated alditol, deriving from the uronic acid, from the non-deuterated analogue obtained from the corresponding neutral sugar. Subsequent identification of the methylated sugar derivatives is usually accomplished through GLC–mass spectroscopy.

Other information on the sequence of monosaccharide linkages can be obtained through periodate oxidation, Smith degradation, or techniques involving β-elimination. The Smith degradation procedure involves periodate oxidation of vicinal hydroxy groups, followed by reduction of the product with borohydride to form a polyalcohol. The bonds in this product are then hydrolysed under relatively mild conditions. The resultant oligomeric or low molecular mass glycosides can then be identified. Steps used in identification of the structure of xanthan can be seen in Figs 2.2 and 2.3. When amino sugars are present they are usually deacetylated by treatment with a suitable base. Deamination may also be used.

FAB–mass spectrometry

Fast atom bombardment-mass spectrometry (FAB–MS) has considerably extended the potential for studying polysaccharide fragments. Mass spectra with both positive and negative ions can be obtained; these, together with the appearance of pseudomolecular ions (e.g. $[M + Na]^{-1}$ $[-M-H]^{-}$) can provide information on molecular mass. The information from fragment ions provides a valuable means of establishing the sequence of monosaccharides and non-carbohydrate components in linear polysaccharides. When complemented by the production of specific fragments

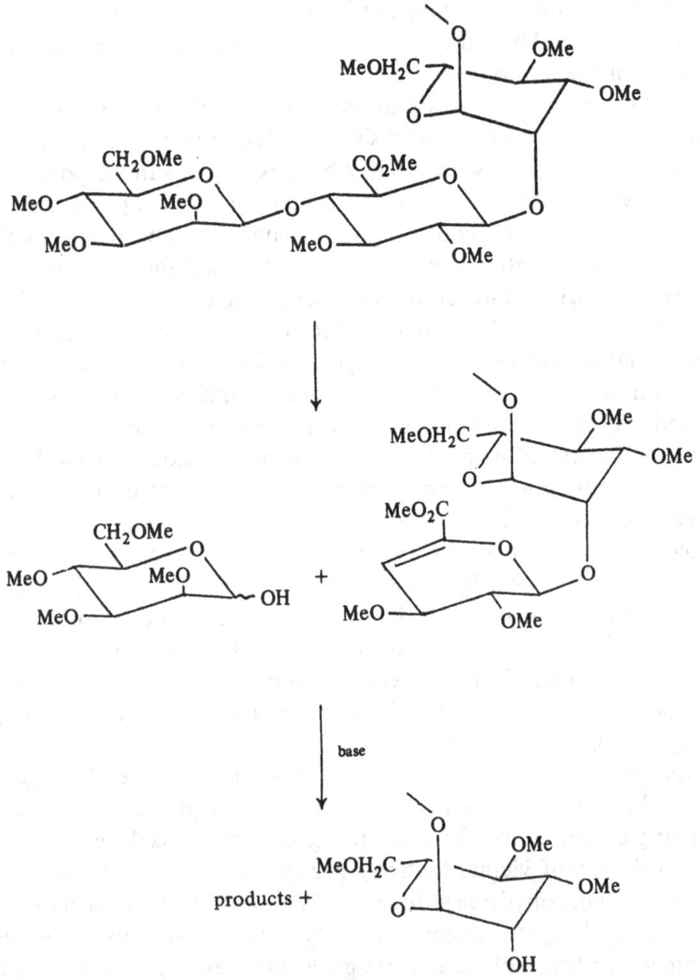

Fig. 2.2. Use of base-catalysed β-elimination in determining the structure of xanthan. (Reproduced from Lindberg (1981) with permission.)

through the use of specific degradative enzymes, FAB–MS provides a powerful tool for deducing polysaccharide fine structure.

A further use for FAB–MS is to determine the position of O-acylated residues in the oligosaccharides obtained from hydrolysis by endoglyca-nases. FAB–MS provides a very useful technique for analysing polar molecules to which labile functional groups are attached. No derivatisation or volatilisation of the sample is needed. It is possible to obtain exact molecular mass information and thus determine the nature and the number of substituent groups that are present. The oligosaccharides can also be

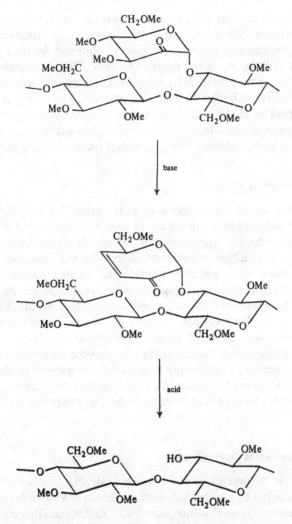

Fig. 2.3. Degradation of methylated xanthan. (Reproduced from Lindberg (1981) with permission.)

permethylated, or they may be per-*O*-acetylated by using perdeuteroacetic acid under conditions which do not remove the pre-existing acyl groups. The product is then subjected to FAB–MS to yield data from which the composition and sequence of the oligosaccharides can be determined.

Oligosaccharides

The regular structures found in most microbial polysaccharides can frequently be hydrolysed by highly specific enzymes, although these are seldom available commercially (Chapter 4). The resultant oligosaccharides can then be analysed by mass spectroscopy or fast atom bombardment spectroscopy. Considerable information can be obtained from oligosaccharides as well as from the initial polysaccharides. The oligosaccharides can be obtained from the polymers by partial acid hydrolysis or by the use of specific enzymes (see Chapter 4). They can be purified by gel permeation chromatography or by HPLC, then analysed by using the same procedures as for intact polysaccharides. In addition, various glycosidases may be employed to remove and to identify terminal sugars on the oligosaccharides.

Immunological methods

Some information about the structure of polysaccharides and oligosaccharides may be obtained through the use of specific antisera (and lectins). The terminal non-reducing sugars, especially those of polysaccharide sidechains, together with adjacent residues, determine the immunological specificity. Provided that suitable antisera to known carbohydrate structures are readily available, serological methods can indicate the likely structure of the terminal portions of the polysaccharide molecule. The use of monoclonal antibodies may also provide specificity towards acylated residues such as 3,4-pyruvylated or 4,6-pyruvylated D-galactose from *Klebsiella* polysaccharides or the terminal pyruvylated D-mannose residues in xanthan. The actual immunological techniques employed to demonstrate cross-reactivity may be important, as tests using erythrocytes coated with polysaccharides have proved to be more sensitive than immunodiffusion tests.

2.4. Molecular mass determination

It is now possible to determine the relative molecular masses (M_r) of polysaccharides, either by comparison with standards of known M_r having similar conformation, or by absolute procedures. Gel permeation chromatography can be applied in HPLC systems but suffers from a number of problems including sample viscosity and lack of suitable standards. Pullulan fractions of relatively well-defined size are available and permit fairly accurate determination of relative molecular masses of less than one million. This is satisfactory for some microbial polysaccharides, but most are of higher molecular mass.

An alternative procedure for the determination of relative molecular masses is the use of values obtained from instrinsic viscosity measurements employing the Mark Houwink equation. If this technique is used, it is probably necessary to ensure that all measurements are made using solutions in the same molar concentration of NaCl. Random coil polyelectrolytes can be expected to be significantly affected at different salt

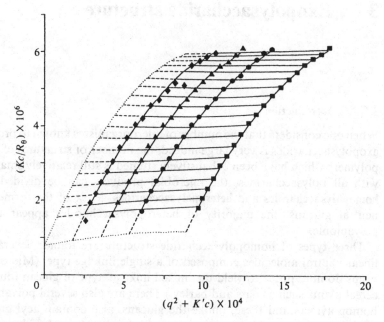

Fig. 2.4. A static Zimm plot of the light-scattering measurements for a xanthan preparation in 0.1 M KCl solution.

concentrations, and polysaccharides such as xanthan, with its tendency to form aggregates in the presence of salts, may give erroneous results.

To obtain accurate measurement of relative molecular mass without reliance on standards, light scattering procedures can be used. Provided that the polysaccharide solutions are clear and free from suspended particulate matter and aggregates, two techniques are available. In the Zimm plot, the more useful technique, values are extrapolated from the intercept at zero scattering angle and zero concentration in solutions in which the intensity of light scattered is measured as a function of angle. An example of such a plot is shown in Fig. 2.4.

Further reading

Aspinall, G. O. (1982). *The polysaccharides*, vol. 1. Academic Press, New York and London.

Chaplin, M. F. and Kennedy, J. F. (1986). *Carbohydrate analysis. A practical approach*. IRL Press, Oxford and Washington.

Lindberg, B. (1981). Structural studies of polysaccharides. *Chemical Society Reviews* **10**, 409–34.

Whistler, R. L. and BeMiller, J. N. (1980). *Methods in carbohydrate chemistry*, vol. 8. Academic Press, New York and London. [Also earlier volumes in this series.]

3 Exopolysaccharide structure

3.1. Introduction

When one considers that the number of microorganisms known to produce exopolysaccharides is very large indeed, the number of structures of these polymers which have been exhaustively studied is still relatively small. As with all polysaccharides, the microbial products can be divided into **homopolysaccharides** and **heteropolysaccharides**. Most of the former are neutral glucans; the majority of heteropolysaccharides appear to be polyanionic.

Three types of homopolysaccharide structure are found. Several are linear neutral molecules composed of a single linkage type. (Microorganisms do not appear to yield the 'mixed linkage' type of glucan found in cereal plants such as oats and barley.) There are also several polyanionic homopolymers and these, unlike the glucans, also contain acyl groups. Slightly more complex structures are the homopolysaccharides of the **scleroglucan** type, which possess tetrasaccharide repeating units due to the 1,6-α-D-glucosyl side-chains present on every third main chain residue. Finally, branched homopolysaccharide structures are found in dextrans.

Microbial heteropolysaccharides are almost all composed of repeating units varying in size from disaccharides to octasaccharides. These frequently contain one mole of a uronic acid, which is usually D-glucuronic acid. Very occasionally, two uronic acids are present. The uniformity of the repeat units is based on chemical studies and it is *possible* that some irregularities may be found, especially in the polymers composed of larger and more complex repeat units. The heteropolymers commonly possess short side-chains, which may vary from one to four sugars in length. Very rarely, the side-chains themselves may also be branched. It is also possible to find exopolysaccharides that contain several different side-chains, usually in the form of single monosaccharides attached to adjacent main chain residues. Some bacterial alginates are exceptional in being heteropolysaccharides of irregular structure.

3.2. Homopolysaccharides

β-D-glucans

Cellulose

Although **cellulose** is normally thought of as a polymer from the walls of eukaryotic plants, slime moulds and algae, it is also produced by

20

(a) — — — -β-D-Glc *p*-1, 3-β-D-Glc *p*-1, 3-β-D-Glc *p*- — — —

(b) — — — -β-D-Glc *p*-1, 3-β-D-Glc *p*-1, 3-β-D-Glc *p*- — — —
$$\uparrow 6$$
$$1|$$
β-D-Glc *p*

Fig. 3.1. The structure of curdlan (a) and of scleroglucan (b).

Acetobacter xylinum and possibly by a range of other, mainly Gram-negative, bacterial species. In these bacteria, cellulose is an exopolysaccharide that is excreted into the medium, where it rapidly aggregates as microfibrils. Although production of cellulose from bacteria is not normally a commercial process, it does provide a source of pure polymer free from lignin and other related material. This may be useful for experimental purposes and also has the advantage that it can readily be prepared in an isotopically labelled form. The bacteria also provide a valuable system for the study of cellulose biosynthesis.

Curdlan
A number of bacterial strains, including *Alcaligenes faecalis* var. *myxogenes*, *Agrobacterium* species and *Rhizobium* species, each produce several exopolysaccharides. One of these is a neutral gel-forming glucan of simple structure and relatively low molecular mass (*ca.* 74000). The polymer has been named **curdlan** and is a 1,3-β-D-glucan (Fig. 3.1) which is insoluble in water.

Scleroglucan
Scleroglucan and related β-D-glucans are produced by several fungal species, including *Sclerotium rolfsii* and the wood-rotting basidiomycete *Schizophyllum commune*. Despite their close structural similarity to curdlan, they are soluble polysaccharides. They possess a 1,3-β-D-linked main chain of glucose residues, attached to which are 1,6-β-D-glucosyl residues (Fig. 3.1). The side-chains are probably attached regularly on every third glucose in the main chain. Commercial polymers have a relative molecular mass of about 1.3×10^5. Some of the glucans, such as those from *Sclerotium glucanicum*, are of lower molecular mass (*ca.* 18000); substitution on some of these β-D-glucans may be on every fourth or even every sixth main chain unit (Table 3.1). The fungal β-D-glucans of this type comprise a large group of homopolysaccharides which vary somewhat in the number and length of their side-chains. These may be single glucosyl residues as indicated for scleroglucan, or they may be longer; it is also not clear whether they are always regularly distributed on the main chain or whether distribution is random.

Table 3.1. *Fungal β-glucans*

Source	M_r (where known)	Degree of branching[a]
Aureobasidium pullulans	10^5–10^6	10–66%
Auricularia auricula-judae	—	66%
Claviceps sp. 47A	—	25%
Claviceps fusiformis	—	33%
Lentinus edodes	—	40%
Monolinia fructigena	7×10^4	50%
Piricularia oryzae	—	17%
Porio cocos	3.7×10^5	very low
Pythium acanthicum	—	33%
Sclerotium glucanicum	1.8×10^4	33%
Schizophyllum commune	—	25–33%

Note:
[a] Percentage of main-chain residues carrying glucosyl side-chains.

α-D-glucans

Dextrans
Several bacterial species produce **dextrans**, exopolysaccharides composed entirely of α-linked D-glucosyl residues. All such polymers are predominantly (1→6) linked. However, different types of dextran are recognised. In some, there may be almost no other type of linkage; alternatively, up to 50% of the glucose residues may be linked 1,2; 1,3 or 1,4. Dextrans vary considerably: some contain one type of linkage only in addition to 1,6-α (as is found in the normal commercial product), while others may be composed of three or even four linkage types. Most dextrans are polymers of high molecular mass. Industrial dextran production is mainly from a strain of *Leuconostoc mesenteroides*, which yields a polysaccharide with about 95% 1,6 linkages and 5% 1,3 linkages and with a relative molecular mass of about 4–5 × 10^7 (Fig. 3.2).

Elsinan
A further α-D-glucan, **elsinan**, obtained from culture filtrates of *Elsinoe leucospila*, has some structural similarity to pullulan, as well as to the polysaccharide nigeran from *Aspergillus niger*. It is a linear polysaccharide composed mainly of maltotriose units linked 1,3-α to each other, but also containing a small number of similarly linked maltotetraose units. Enzymes have proved extremely useful in determining the structural details of this polymer; in particular, they have provided evidence for the presence of sequences of 10–11 maltotriose units. It is thus composed of similar structural units to pullulan (which also contains a small number of maltotetraose units) but the linkage is 1,3 instead of 1,6 as in pullulan. Although **nigeran** also contains 1,3-α-D-glucosyl linkages, these alternate

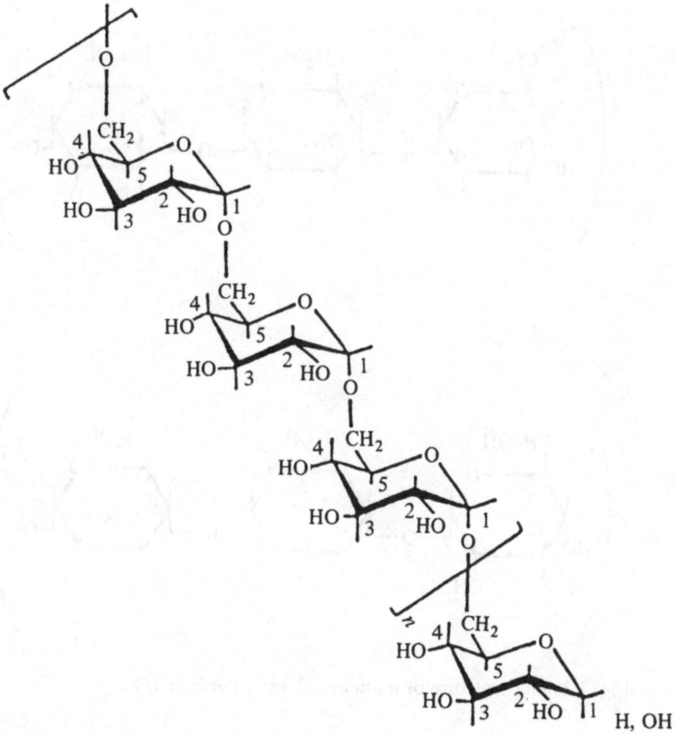

Fig. 3.2. Typical dextran structure.

on a regular basis with 1,4-linkages. Elsinan is readily soluble in water, giving highly viscous solutions. At higher concentrations, gels are formed.

Pullulan

The exopolysaccharide **pullulan**, from the dimorphic fungus *Aureobasidium pullulans*, has received considerable attention. It is composed of maltotriosyl repeat units linked 1,6-α to form a polymer of relative molecular mass 10^5–10^6. Very similar products are formed by other fungal species, including *Tremella mesenterica* and *Cyttaria harioti*. The structure of pullulan (Fig. 3.3) resembles that of elsinan, in that each represents a homopolysaccharide with a regular repeat unit (maltotriose). In both polysaccharides, there are small numbers of maltotetraose units in place of the trisaccharides. The distribution of the maltotetraose units in pullulan appears to be irregular, but it is a linear polymer, lacking any side-chains.

Sialic acids

Strains of *Escherichia coli* and of *Neisseria meningitidis* secrete **sialic acid** exopolysaccharides. Group C meningococcal sialic acid is a homopolysac-

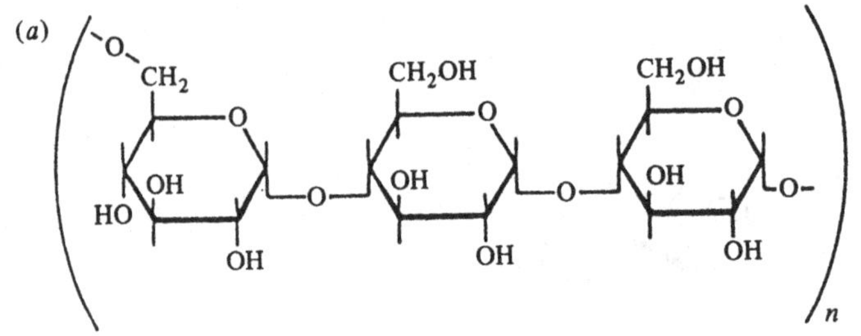

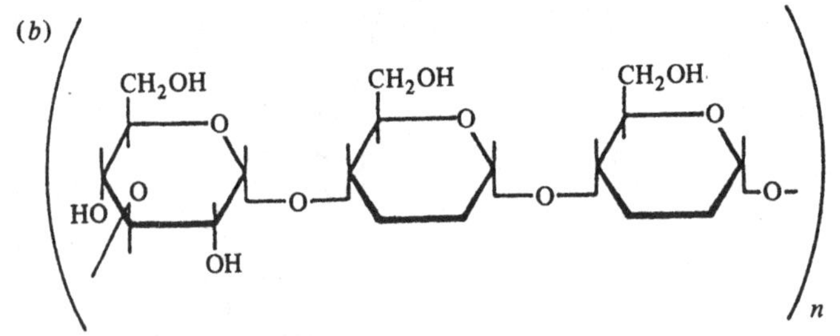

Fig. 3.3. The structure of pullulan (a) and of elsinan (b).

charide in which the residues are $(2\rightarrow9)$ linked. The products from *E. coli* K1 and *N. meningitidis* group B are $(2\rightarrow8)$-α-linked. These polymers are susceptible to degradation with a phage-induced enzyme, whereas the 2,9-linked polysaccharides are resistant. Another bacterial sialic acid (from several other different strains of *E. coli*) is non-acetylated and is composed of *alternate* 2,8-α- and 2,9-α-linkages. It, too, is susceptible to hydrolysis by the phage-induced enzymes. In addition to these homopolymers, some strains of *N. meningitidis* synthesise heteropolysaccharides containing both sialic acid and D-glucose (Fig. 3.4). Several of the bacterial sialic acids possess hydrophobic 1,2-diacylglycerol groups at the polysaccharide chain termini, causing the polymers to aggregate in the form of micelles. This may also provide a mechanism for attaching the capsular polysaccharides to the bacterial surface.

3.3.   Heteropolysaccharides

Bacterial alginates

Commercially available **alginates** are currently isolated from marine algae such as *Laminaria* and *Microcystus*. The discovery of bacterial alginates, polymers of irregular structure also composed of D-mannuronic acid and L-guluronic acid, resulted in several proposals that they could substitute for

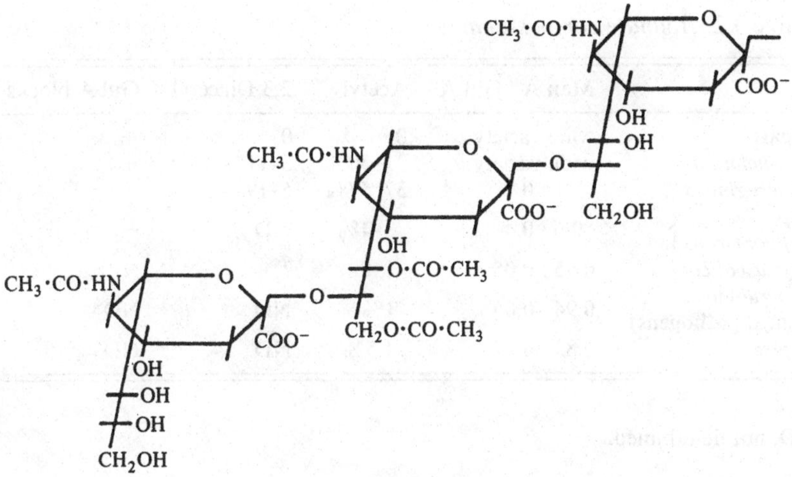

Fig. 3.4. The structure of a bacterial sialic acid, the K1 antigen of *E. coli*. Other sialic-acid-containing exopolysaccharides are 2,8-α-linked, 2,9-α-linked or alternatively 2,8- and 2,9-α-linked. Others also have D-glucose or D-galactose residues. The central *N*-acetylneuraminic acid residue is shown *O*-acetylated at C7 and C9, as is found in form variants of this polysaccharide.

the algal products. Further structural investigations on alginates from the range of bacterial species now known to synthesise this type of polysaccharide, has revealed a range of different polymer types, all with similar chemical composition. All are composed of the same two uronic acids as algal alginate, but in addition, many of them are highly acetylated, the acetyl groups being carried solely on D-mannuronosyl residues in the polymers. As can be seen from the details presented in Table 3.2, bacterial alginates are a family of exopolysaccharides which, despite their similar composition, vary considerably in their structure. The material from *Azotobacter vinelandii* bears the closest resemblance to algal alginate, although unlike the latter, it contains *O*-acetyl groups. Both are composed of three types of structure: poly-D-mannuronic acid sequences; poly-L-guluronic acid sequences; and mixed sequences (Fig. 3.5). These are sometimes termed 'block structures'. *Azotobacter chroococcum* synthesises two polysaccharides, one of which is an acetylated alginate with high D-mannuronic acid content. The majority of *Pseudomonas* species that synthesise alginate also yield exopolysaccharides with relatively high mannuronic acid content. The interesting feature of the *Pseudomonas* products is that, unlike the algal or *Azotobacter* alginate, there are no contiguous sequences of L-guluronic acid residues. Whereas the properties of *A. vinelandii* polysaccharide can therefore be expected to have much in common with the commercial algal products, the others are quite different polysaccharides. The alginates from the plant pathogenic *Pseudomonas*

Table 3.2. *Alginate composition*

	Man A : Gul A	Acetyl	2,3 Diacetyl	Gul A blocks
Algal	wide variety	0	0	+
A. *vinelandii*	wide variety	21–50%	3–11%	+
P. *aeruginosa*	1 : 0	37–57%	5–19%	−
P. *putida* P. *fluorescens*	0.6 : 0.4	3–4%	ND	−
P. *phaseolicola*	0.95 : 0.05	1%	?	−
P. *aeruginosa* (fungal pathogens)	0.94 : 0.06	1%	ND	ND
P. *pisi*	0.83 : 0.17	4.5%	ND	ND

Note:
ND, not determined.

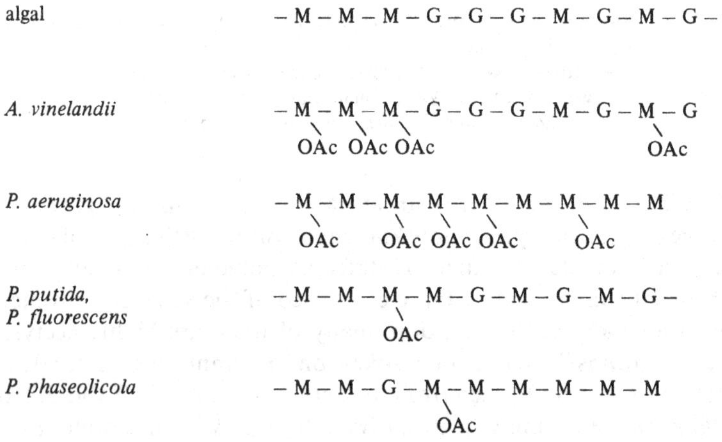

Fig. 3.5. Bacterial and algal alginate structures. M, mannuronic acid; G, guluronic acid; OAc, acetyl ester. Esters may be on C_2, C_3, or both.

species are also generally of lower molecular mass and higher polydispersity. Unlike *P. aeruginosa* material, they mainly have relatively low acetyl content. In the highly acetylated polysaccharides from either *A. vinelandii* or *P. aeruginosa*, some D-mannuronosyl residues may carry *O*-acetyl groups on the C_2 and C_3 positions, i.e. between 3 and 11% of these residues carry acetylation on both carbon atoms.

Emulsan and related polysaccharides

Acinetobacter calcoaceticus is a capsulate Gram-negative bacterium, the exopolysaccharides from which are potent emulsifying agents. Structural

→3)-α-L-Rha-(1→3)-α-D-Man-(1→3)-α-L-Rha-(1→3)-α-L-Rha-(1→3)-β-D-Glc-(1→
$\qquad$ 2
$\qquad$ ↑
$\qquad$ 1
$\quad$ α-L-Rha-(1→4)-β-D-GlcA

Fig. 3.6. Structure of a bacterial biosurfactant from a strain of *Acinetobacter*:
the polysaccharide from *A. calcoaceticus* strain BD4.

studies have shown that two of these polymers differ considerably in their
chemical structure, despite their common functional properties. Strain
BD4 polysaccharide is a heptasaccharide repeat unit composed of
rhamnose, mannose, glucose and glucuronic acid. Hydrolysis with an endo-
β-glucosidase, derived from bacteriophage, yielded the heptasaccharide
(Fig. 3.6). **Emulsan** from strain RAG-1 contains D-galactosamine, an
aminouronic acid and an amino sugar. It also contains 15% fatty acyl *O*-
esters. The repeating unit is thus smaller than that of BD4 polysaccharide
and BD4 lacks both amino sugars and *O*-esters.

Gellan and related polymers

The Kelco Division of Merck in San Diego has maintained their sequence
of new microbial polysaccharide introductions, first with **gellan**, then with a
series of structurally related polysaccharides. In its native form, gellan
carries both *O*-acetyl and glyceryl substituents on a linear polymer
composed of tetrasaccharide repeat units. Subsequently, further polysaccharides with the same main-chain sequence were obtained from other
bacterial strains. However, these carried differing side-chains: a rhamnose-
containing or a glucose-containing (gentibiosyl) disaccharide or, in one
polysaccharide either L-rhamnose or L-*mannose*. This latter polysaccharide
is thus highly unusual in containing (i) L-mannose and (ii) a variable side-
chain. The ratio of L-rhamnose to L-mannose is approximately 2 : 1. Two
further polymers were even more unusual in that, in both, one of the *main-
chain* sugars varied. It could again be either L-mannose or L-rhamnose. One
of the exopolysaccharides (S88) also contained 5% acetate. The structures
proposed for this series of polysaccharides are shown in Fig. 3.7.

Hyaluronic acid and heparin

Although no microbial strains produce heparin, a strain of *Escherichia coli*
serotype K5 does form a capsular polysaccharide in which the disaccharide
repeat unit is essentially a form of **desulphatoheparin**. The polymer is
composed of a repeating unit of 4-β-D-glucuronosyl-1,4-α-*N*-acetyl-D-
glucosamine. In contrast to this, desulphatoheparin from eukaryotic
sources is normally composed of alternate disaccharides containing D-
glucuronosyl amino sugar and L-iduronosyl-1,4-α-*N*-acetyl-D-glucosamine.

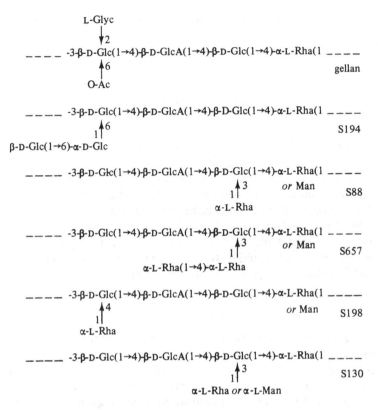

Fig. 3.7. The structures of gellan from *Pseudomonas elodea* and of related polysaccharides.

The bacterial product most closely resembles type II glycosaminoglycuronan chains. These are synthesised in eukaryotes in the Golgi complex and then polymerised onto core proteins. Subsequently, they may be modified (i.e. after polymerisation) through the action of uronosyl epimerase and sulphatotransferases. These enzymes introduce the L-iduronosyl and sulphate residues, respectively, both of which are absent from the bacterial product. Because of its structural similarity to heparin, this bacterial polysaccharide may well have some potential in medical research and in determination of the specificity of heparinases and related enzymes. Another interesting polymer of this general type from *E. coli* K4, possesses a chondroitin backbone to which β-D-fructofuranosyl residues are attached at the C_3 position of the D-glucuronic acid. After removal of the fructosyl residues by mild acid treatment, the polysaccharide is a substrate for both hyaluronidase and chondroitinase.

Several bacterial species are capable of producing **hyaluronic acid**, apparently identical in chemical structure to that obtained from eukaryotic

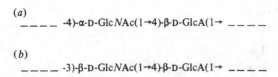

(a)

____ -4)-α-D-GlcNAc(1→4)-β-D-GlcA(1→ ____

(b)

____ -3)-β-D-GlcNAc(1→4)-β-D-GlcA(1→ ____

Fig. 3.8. The structure of the desulphatoheparin-like polymer from *E. coli* K5 (a) and of bacterial hyaluronic acid (b).

material. The polysaccharide is composed of repeating units of 1,4-β-linked disaccharides of D-glucuronosyl-1,3-β-N-acetyl-D-glucosamine. The bacterial sources of hyaluronic acid are *Pseudomonas aeruginosa* strains and group A and group C streptococci. The product from one group C streptococcal strain forms two distinct fractions, a high molecular mass cell-bound product and a soluble exopolysaccharide of average relative molecular mass 2×10^6. It was suggested that the cell-bound polymer of initial relative molecular mass 10×10^6 was hydrolysed by a membrane-bound hyaluronidase and then released in a soluble form. Material from group A streptococci, obtained in a yield of 0.5–1.0 g l^{-1}, was relatively pure hyaluronic acid, over 91% of which was accounted for as amino sugar and uronic acid. Although bacterial production of hyaluronic acid has been commercialised, the nature of the bacterial strain and the size of the material produced have not been reported. The bacterial product is identical in structure with hyaluronic acid from vertebrate species and is free from protein. In eukaryotic material, it is found in many tissues; in one of these, the vitreous humour of the eye, it is found as a non-sulphated polymer unlinked to protein as is true of the bacterial exopolysaccharide (Fig. 3.8).

Rhizobium *heteroglycans*

Some species of *Rhizobium*, namely *R. trifolii*, *R. meliloti* and *R. leguminosarum*, but **not** *R. phaseoli*, form an insoluble neutral capsular polysaccharide. Like curdlan, it is insoluble at room temperature. It is a heteropolysaccharide composed of D-glucose, D-galactose and D-mannose in the molar ratio 1 : 3 : 2, forming a hexasaccharide repeat unit (Fig. 3.9). This polymer is produced as well as soluble polyanionic material, the relative proportions of the two being dependent on the strain and on the growth conditions used. The main chain of the gelling polysaccharide comprises a sequence of α-D-glucosyl-1,3-α-D-mannosyl-1,3-β-D-galactose. The glucose residue carries two side-chains, one of a single D-galactosyl unit and the other a disaccharide composed of D-galactose. The structure thus has a number of unusual features, being a neutral heteropolymer with one main-chain residue carrying two side-chains.

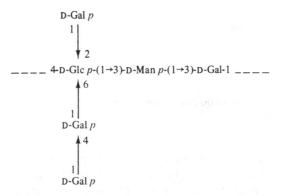

Fig. 3.9. The neutral gel-forming polysaccharide from *Rhizobium meliloti*. (From Zevenhuizen and van Neervan, 1983.)

Succinoglycan

Most of the bacterial strains that yield curdlan also produce **succinoglycan**. As with other examples of microorganisms which can synthesise more than one polysaccharide, strain selection and choice of growth conditions can ensure high yields of one specific polymer. Succinoglycan differs from curdlan in being water-soluble and is composed of octasaccharide repeat units thus conforming to a pattern found in a number of other exopolysaccharides from *Rhizobium* species. There is close structural similarity to the other *Rhizobium* polysaccharides, which all possess highly conserved structures composed of octasaccharide repeat units (see p.5). The succinoglycan is composed of D-glucose and D-galactose in the molar ratio 7:1. Attached to the sugars are three different acyl groups: acetate and succinate esters and pyruvate ketals (Fig. 3.10). The pyruvate is normally present in stoichiometric amounts; the molar ratios of acetate and succinate are commonly of the order of 0.2 and 0.4–0.5 respectively. The monosaccharide components are of course neutral, but the pyruvate ketal situated on the side-chain terminus (D-glucose) confers anionic properties on the polysaccharide. Despite its 1,3-,1,4- and 1,6-linked β-D-glucosyl residues, succinoglycan is resistant to the action of endo-β-glucanases available commercially. However, certain highly specific enzymes have been of great value in elucidating the complex repeat unit structure of this and structurally related polysaccharides (p.5). These enzymes also act on other *Rhizobium* polysaccharides that share some structural features with succinoglycan. Some of them are very highly acylated, carrying 3-hydroxybutyrate on 80% of the galactose residues and two acetyl groups per repeat octasaccharide on main-chain D-glucosyl residues. Other polysaccharides from different strains contained lower levels of 3-hydroxybutyrate and acetate (Philip-Hollingsworth *et al.*, 1989).

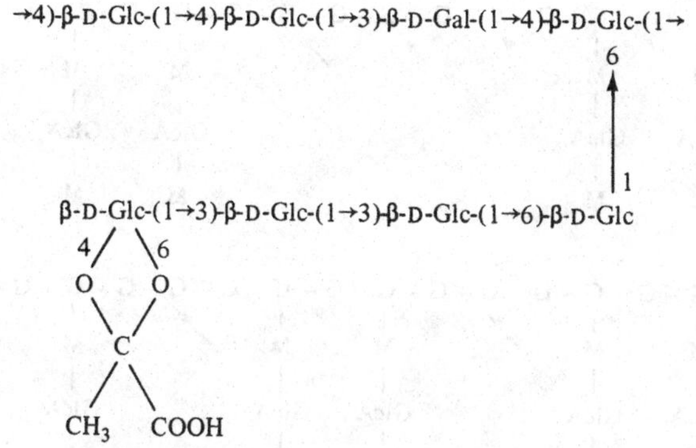

→4)-β-D-Glc-(1→4)-β-D-Glc-(1→3)-β-D-Gal-(1→4)-β-D-Glc-(1→
6
1
β-D-Glc-(1→3)-β-D-Glc-(1→3)-β-D-Glc-(1→6)-β-D-Glc
4
6
O
O
C
CH3
COOH

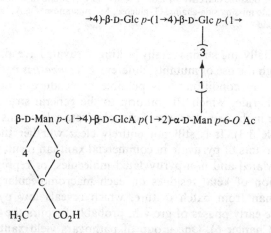

Fig. 3.10. The structure of succinoglycan.

→4)-β-D-Glc p-(1→4)-β-D-Glc p-(1→
3
1
β-D-Man p-(1→4)-β-D-GlcA p(1→2)-α-D-Man p-6-O Ac
4
6
C
H3C
CO2H

Fig. 3.11. The structure of xanthan from *Xanthomonas campestris* pv. *campestris*.

Xanthan

Definitive studies on the structure of **xanthan** showed that it is composed of pentasaccharide repeat units which are of particular interest as they reveal a cellulosic backbone (Fig. 3.11). Alternate glucose residues in the backbone carry trisaccharide side-chains composed of D-mannose and D-glucuronic acid. There have been suggestions that imperfections in side-chain continuity may occur, leaving portions of the backbone either 'bare' or deficient in side-chains, with consequent effects on the physical properties of the polysaccharide (Fig. 3.12). Most commercial xanthan preparations are fully acetylated on the internal D-mannose residue, but only carry pyruvate ketals on about 30% of the side-chain terminal mannose residues. Recently, xanthan with much higher pyruvate content has become

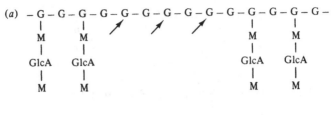

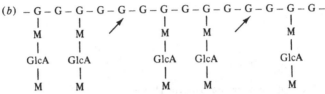

Fig. 3.12. The possible defective structures which may be found in xanthan; arrows indicate missing side-chains. G. glucose; M, D-mannose; GlcA, glucuronic acid.

available commercially and strains totally lacking pyruvate have also been developed. Through the use of mutants, different *X. campestris* pathovars and different nutrient conditions it is possible to produce a range of different polysaccharides which all conform to the general structure of xanthan, but differ in the completeness of carbohydrate side-chains and of acyl groups (Table 3.3). It is still not entirely clear whether the non-stoichiometric amounts of pyruvate in commercial xanthan result from a mixture of pyruvylated and non-pyruvylated molecules, or represents a random distribution of ketal residues on each macromolecular chain. Analysis of xanthan from batch culture, which reveals low pyruvate content during the early phases of growth, probably favours the former explanation (see Chapter 6). One group of pathovars yield xanthan in which the content of pyruvate is very low, but in which the internal mannosyl residue in the side-chain carries *two* moles of acetate. Following lengthy genetic studies on xanthan production, mutant polymers lacking the terminal D-mannose (and pyruvate) or lacking the terminal **disaccharide** are also available. The latter naturally differs from xanthan in being a neutral polymer, although it still possesses useful rheological properties. The various types of repeating unit to be found in different xanthan preparations are illustrated in Fig. 3.13.

As well as xanthan from wild type or mutant strains of *X. campestris* pv. *campestris* and related strains, several other polysaccharides have been shown to share part of the structure and also possess exploitable properties. The ability of mutant strains to produce 'polytrimer' and 'polytetramer' (i.e. xanthan molecules lacking the side-chain disaccharide and monosaccharide termini, respectively) has been mentioned. In addition to these variants, several strains of *Acetobacter xylinum* yield apparently xanthan-

Table 3.3. *Types of xanthan structure*

		Molar ratio			
Mannose	Glucose	Glucuronic acid	Acetate	Pyruvate	
2	2	1	1	0.3	Most commercial products
2	2	1	1	0	Mutant strain developed commercially
2	2	1	<0.1	<0.1	X. phaseoli strain
2	2	1		0.3	X. phaseoli strain
2	2	1	1	~0.7	Commercial product
2	2	1	~2	~0.1	Certain strains and pathovars
2	2	1	0	0	Can be prepared chemically
2	2	1	1	0.3	'Low viscosity' material (patent)
<2	2	<1	0	0	Mutant preparations

(a) – G – G –

(b) – G – G –   – G – G –
      | |
      M   M – Ac

(c) – G – G –   – G – G –
      | |
      M   M – Ac
      | |
    GlcA   GlcA

(d) – G – G –   – G – G –   – G – G –   – G – G –   – G – G
      | | | | |
      M   M   M – Ac   M – Ac Ac – M – Ac
      | | | | |
   GlcA   GlcA   GlcA   GlcA   GlcA
      | | | | |
      M   M = Pyr   M   M = Pyr   M

Fig. 3.13. The various repeat units occurring in preparations of xanthan from wild-type (*d*) and mutant (*a–c*) bacteria. M, D-mannose; G, D-glucose; Ac, acetyl; Pyr, pyruvate.

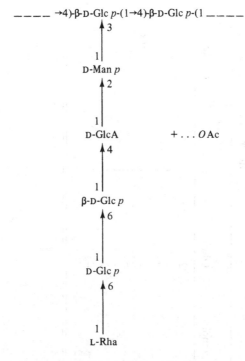

_ _ _ _ _ →4)-β-D-Glc *p*-(1→4)-β-D-Glc *p*-(1 _ _ _ _ _
                                 ↑3
                                 1
                            D-Man *p*
                                 ↑2
                               1
                        D-GlcA   + ... *O*Ac
                                 ↑4
                               1
                          β-D-Glc *p*
                                 ↑6
                             1
                          D-Glc *p*
                                 ↑6
                             1
                          L-Rha

Fig. 3.14. The structure of acetan from *Acetobacter xylinum*. (From Tayama *et al.*, 1985.)

$$\text{---- } \rightarrow 4)\text{-D-GlcA-}(1\overset{\alpha}{\rightarrow}3)\text{-L-Fuc-}(1\overset{\alpha}{\rightarrow}3)\text{-D-Glc}(1\text{- ----}$$

$$\uparrow 4$$

$$1|$$

$$\beta\text{-D-Glc}$$

Fig. 3.15. The structure of the polymer from *Enterobacter aerogenes* strain XM6.

like polysaccharides. As bacterial cellulose is a normal product of *A. xylinum*, the ability to modify cellulose by the addition of side-chains is perhaps not unexpected. One product has been characterised as having a cellulosic main-chain together with a *pentasaccharide* side-chain on alternate main-chain sugars. This polymer closely resembled xanthan, the terminal β-D-mannosyl residue being replaced by an L-rhamnosyl-gentibiosyl sequence (Fig. 3.14). Like xanthan, this polysaccharide is acetylated. Another strain of the same bacterial species is thought to synthesise a polysaccharide which differs only in the conformation of one of the side-chain linkages. These xanthan-like polymers are produced in addition to cellulose in some strains or replacing it in others. They form highly viscous aqueous solutions but may not be as potentially valuable as xanthan. Another polysaccharide of the *same composition*, but apparently different structure, has also been found in a cellulose-negative *A. xylinum* strain. In a further strain of the same species, AM1, the polysaccharide is unusual in that the repeat unit is claimed to be large (11 monosaccharides) and the cellulosic backbone structure carries two very dissimilar side-chains separated from each other by an unsubstituted D-glucose residue. Each side-chain is attached at the 3-position of a main-chain D-glucose residue.

XM6

A polysaccharide with interesting gelation properties is produced by an *Enterobacter* strain (XM6); this polymer proved to resemble very closely polysaccharides from *Klebsiella aerogenes* type 54. Each polysaccharide was composed of the same tetrasaccharide repeat unit (Fig. 3.15). However, the *Klebsiella* exopolysaccharides carry *O*-acetyl groups either on every L-fucose residue or on **alternate** ones, whereas the XM6 exopolysaccharide is devoid of acyl substituents. These aspects of fine structure were readily confirmed through degradation with specific enzymes obtained from bacteriophages.

Agrobacterium, Rhizobium *and* Zoogloea *polysaccharides and related polymers*

Floc-forming bacteria play an important role in the purification of activated sludge or water with a high content of organic material. In the

Table 3.4. *Galactoglucopolysaccharides*

Source	Glucose	Galactose	Pyruvate	Acetate	Succinate
Achromobacter sp.	1	1*	0.81–0.99	—	—
Agrobacterium radiobacter	0.9	1	0.83	—	—
Pseudomonas marginalis	1	1*	1	—	1
Rhizobium meliloti	1	1	0.9	—	—
Rhizobium sp.	7	1	1	—	—
Zoogloea sp.	7	1	—	—	—

Note:

* = pyruvylated sugar

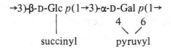

→3)-β-D-Glc p(1→3)-α-D-Gal p(1→

Fig. 3.16. The structure of marginalin, recently isolated from *Pseudomonas fluorescens* strains.

purification process, the exopolysaccharides synthesised by *Zoogloea ramigera* and related species play important roles. Some of these polysaccharides are claimed to have an unusually high affinity for metallic ions and for amino acids. The polymer from one strain has been shown to be composed of D-glucose, D-galactose and pyruvate. It has been suggested that the main chain of this polymer contains 1,4-β-linked D-glucose residues, but there are still conflicting views about its exact structure.

Several other bacterial polysaccharides composed solely of the neutral sugars D-glucose and D-galactose have been reported. The structures of most of these polymers have still not been elucidated, although several appear to contain equimolar amounts of glucose and galactose, together with acyl substituents (Table 3.4). However, one interesting polymer has been isolated from strains of *Pseudomonas marginalis* and shown to be composed of disaccharide repeat units to which pyruvate and succinate groups are attached (Fig. 3.16).

Clearly, the exopolysaccharides of potential industrial interest represent a very varied series of chemical structures. Some do, however, reveal considerable conservation of structure. They are composed of a relatively limited range of monosaccharides and acyl substituents. They also provide opportunities for the correlation of carbohydrate structure with physical properties.

Further reading

Aspinall, G. O. (1983). *The polysaccharides*, vol. 2. Academic Press, New York.

Stivala, S. S., Crescenzi, V. & Dea, I. C. M. (1987). *Industrial polysaccharides. The impact of biotechnology and advanced methodologies.* Gordon and Breach, New York.

4   Enzymes degrading exopolysaccharides

4.1.   Introduction

While microbial exopolysaccharides, in common with similar polymers from other sources, are the substrates for degradative enzymes, the number of polysaccharases that have been isolated and characterised is relatively small. Only a small number of the polysaccharide-producing microbial species also yield enzymes degrading the same polymers. The exceptions include some of the bacterial species synthesising alginate and hyaluronic acid. A rich source of enzymes degrading bacterial exopolysaccharides has proved to be bacteriophages. These viral particles contain polysaccharases as part of the particle structure, usually in the form of small spikes attached to the base-plate of the phage. After phage infection, the bacterial lysates normally contain further amounts of the same enzyme in soluble form. The advantage of bacteriophages as sources of enzymes degrading polysaccharides is their freedom from other associated glycosidases, which might further degrade any oligosaccharide products. On the other hand, yields of phage-induced enzymes are low and they can only be regarded as laboratory tools of value in structural studies, unless the genes for the enzymes can be cloned and expressed on a large scale in microbial hosts. In addition, not all bacteriophages for exopolysaccharide-producing bacteria yield such enzymes.

There are very few commercially available enzymes acting on microbial polysaccharides. Consequently, the laboratory interested in using enzymes for structural determinations or for quality control must normally isolate its own enzymes. A number of polysaccharases have been obtained from bacterial and fungal sources by using enrichment procedures, with the polysaccharides as substrates. Several purified enzymes from such sources have been described, but a major problem is the discovery that, although bacterial consortia frequently show polysaccharide-degrading capacity, the individual bacterial strains do not readily synthesise the specific enzymes desired. A further disadvantage is the presence of a wide range of glycanases and contaminating glycosidases, which may prove difficult to remove during enzyme purification.

Almost all the enzymes capable of microbial polysaccharide degradation show a very high degree of substrate-specificity. They seldom degrade more than one polymer, even when a high degree of homology exists between the polysaccharide structures. Consequently, commercial polysaccharases rarely prove to have enzyme activity against microbial exopolysaccharides.

This is in marked contrast to the range of such enzymes available for degrading commonly used plant or animal polysaccharides and glycoproteins, including starch, cellulose, gluco- and galactomannans and pectin. Polysaccharases provide a valuable adjunct to chemical methods of structural determination; additionally, they permit the production of acylated fragments for study by fast atom bombardment spectrometry (FAB) and related methods, whereas the labile acyl groups are usually lost when chemical methods are involved. They also provide the basis for highly specific assay procedures and for determining the uniformity of polysaccharide structure.

Polysaccharide-degrading enzymes are of course also a problem to both the user and the producer of microbial polysaccharides. Their presence in any formulation may destroy the physical characteristics for which the polymer has been employed. Equally, the presence of an enzyme deriving from the microbial culture can affect other components with which the polysaccharide is later mixed. Thus, xanthan preparations must normally be pasteurised to destroy the bacterial enzyme *cellulase* if xanthan is to be mixed with other polymers, such as carboxymethylcellulose, which could be destroyed by the enzyme. The problems of polysaccharase enzyme synthesis in microbial cultures for polysaccharide production are discussed in Chapter 6.

4.2. Hydrolytic enzymes (polysaccharases)

β-D-glucanases

Cellulase
Although cellulose is normally obtained as a polymer from plant sources, the fact that it is also produced by *Acetobacter xylinum* means that the bacterial material is also a substrate for *cellulases*. The enzymes are available commercially from several sources including *Trichoderma viride*, *Aspergillus* species and *Bacillus* species. The bacterial polysaccharide provides an excellent substrate for testing the cellulase specificity as it is free of other contaminating polymers.

Scleroglucan hydrolysis
After alkali treatment to yield the random coil form, scleroglucan can be degraded by the 1,3-β-D-glucanase *zymolyase* from *Arthrobacter luteus*. Although only about 50% of the polysaccharide is susceptible to hydrolysis, it can be converted into glucose, laminaribiose and gentibiose (Table 4.1).

α-D-glucanases
Dextranase (E.C. 3.2.1.11) (also 1,3 α-D-glucanase, E.C. 3.2.1.59)
Enzymes degrading dextrans are of relatively common occurrence and are produced by a wide range of microorganisms. These include fungi, Gram-

Table 4.1. *1,3 β-D-glucanases*

Source	Mode of action	Products from	
		Curdlan/laminarin	Scleroglucan
Strongylocentrotus purpuratus	exoglucanase	glucose	—[a]
Saccharomyces cerevisiae	exoglucanase	glucose	—
	endoglucanase	glucose	—
Basidiomycete aphyllophorales	exoglucanase	glucose	—
Bacillus circulans	endoglucanase	glucose, laminaribiose, etc.	—
Penicillium italicum	endoglucanase	glucose, laminaribiose, laminaritriose	—
Rhizopus arrhizus	endoglucanase		glucose, gentiobiose, laminaribiose
Oerskovia xanthinolytica	endoglucanase	laminaripentaose	—
Trichoderma reesei	exoglucanase	glucose	—
Arthrobacter sp.	endoglucanase[b]		

Notes:
[a] Dashes indicate no activity detected.
[b] Products not identified.

Table 4.2. Dextranases

Source	Mode of action	Products
Achromobacter sp.	exohydrolase	isomaltose, limit dextrin
Streptococcus mutans	endodextranase	glucose, isomaltose, isomaltotriose etc.
Pseudomonas sp. UQM733	endodextranase	isomaltotriose, isomaltotetraose,etc.
Cladosporium resinae	endo-1,3-α-D-glucanase	glucose
Pseudomonas sp. NRRL B12324	endo-1,3-α-D-glucanase	isomaltotetraose
Aspergillus nidulans	exoglucanase	glucose
Flavobacterium sp.	endoglucanase	isomaltose, nigerose, nigerotriose, etc.
Arthrobacter globiformis	exo-1,6-α-D-glucanase	glucose
Streptococcus mitis	exo-1,6-α-D-glucanase	glucose

positive bacteria and Gram-negative bacteria (Table 4.2). The enzymes vary widely in their mode of action, some yielding a single product, whereas the majority convert the substrate to glucose and a range of fragments. Because the substrate may vary in the relative proportions of the different glucosidic bonds, enzymes vary in their action on different dextran preparations. Some of the enzymes are **endo**dextranases randomly forming fragments of decreasing size; others are **exo**dextranases removing single D-glucose residues or oligosaccharides successively from the ends of dextran chains.

As well as exodextranases yielding glucose, an extracellular enzyme from an *Achromobacter* species has proved unusual in that it cleaves dextran to produce isomaltose as the sole product. The action of this enzyme resembles that of α-amylase. In addition to the disaccharide, it forms a limit product from the dextran substrate, probably through failure to hydrolyse 1,3-α-branch points in the macromolecule. The limit product can represent about 43% of the original dextran and can be further degraded with a dextranase from *Penicillium luteum*. By analogy with the action of β-amylase on amylopectin (or glycogen), the isomaltose-producing dextranase leaves a macromolecular product from which the outer branches have been removed.

Streptococcus mutans strains which form the 1,3-α-linked dextran **mutan** almost all produce dextranases. These enzymes are **endo**dextranases, which initially yield isomaltose, isomaltotetraose and isomaltopentaose as the major digestion products. Subsequently, the largest of these oligosaccharides is slowly hydrolysed to glucose, isomaltose and isomaltotriose. The enzyme activity is greatest against dextran and decreases with decreasing chain length. By using this type of enzyme, mutan can be converted to isomaltotriose with 72% efficiency. A dextranase from a *Pseudomonas* species also shows some similarity to the *S. mutans* enzymes in its mode of action.

Enzymes with higher specificity have also been proposed for structural studies on dextrans. *Cladosporium resinae* yields an endoenzyme which degrades polymers composed solely of 1,3-α-glucosidic linkages, including the products from *S. mutans*. It does not degrade dextrans from *Leuconostoc mesenteroides* even though they contain up to 40% of 1,3-α-linkages, nor does it give complete hydrolysis of the *S. mutans* products. This was also true of an enzyme from a *Pseudomonas* species; it produced only small amounts of glucose and low molecular mass oligosaccharides, and up to 32% of larger oligosaccharides. Another enzyme from a *Flavobacterium* species hydrolyses both the 1,3-α-linked dextrans and polymers with 1,3 *and* 1,6 linkages, but not soluble dextrans containing solely 1,6-α-linkages.

Clearly, the extent of hydrolysis of dextrans by dextranases and the nature of the products is dependent both on the substrate used and on the

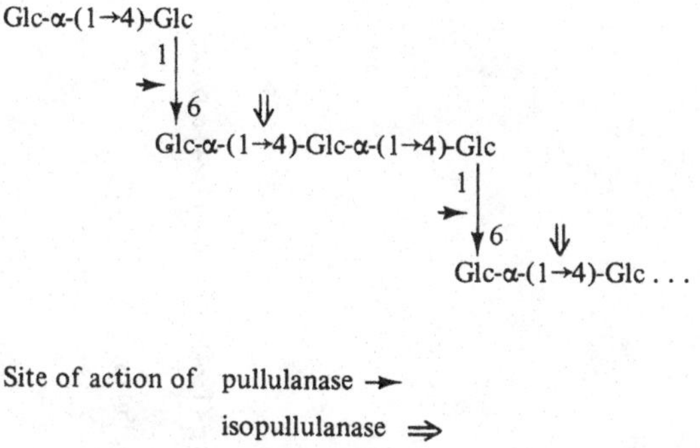

Glc-α-(1→4)-Glc
$$\begin{array}{l} 1 \\ \downarrow 6 \quad \Downarrow \\ \text{Glc-α-(1→4)-Glc-α-(1→4)-Glc} \\ \qquad\qquad\quad 1 \\ \qquad\qquad\quad \downarrow 6 \quad \Downarrow \\ \qquad\qquad\qquad \text{Glc-α-(1→4)-Glc} \ldots \end{array}$$

Site of action of   pullulanase →

isopullulanase ⇒

Fig. 4.1. Sites of action of enzymes degrading pullulan.

enzyme. If the dextrans are highly branched, hydrolysis is much less extensive than is found with linear or slightly branched polysaccharides. The enzymes have the additional value that they can be used to yield relatively large quantities of oligosaccharides of the isomaltose series.

Pullulanase (E.C. 3.2.1.41)

Three specific **endo**enzymes acting on pullulan have been described, each being devoid of effect on the corresponding polysaccharide of uniform linkage. *Pullulanase* cleaves the 1,6-α-D-glucosyl linkage to yield malto-triose as the final product. The enzyme hydrolyses pullulan randomly, the initial products being hexasaccharides and nonasaccharides. These are eventually degraded to give maltotriose. Some activity against the 1,6-α-glucosidic linkages of amylopectin and limit dextrins has also been observed, providing that there are at least two 1,4-α-glycosidic linkages on each side of the bond cleaved. As well as its value in determining the structural details of pullulan, the enzyme provides a convenient means of producing maltotriose in high yield from the substrate. It is also used in combination with glucoamylase to obtain quantitative yields of glucose from partial hydrolysates of starch. It is much less active against glycogen. *Isopullulanase* (E.C.3.2.1.57) from *Aspergillus niger* differs in that it hydrolyses a 1,4-α-glucosidic bond releasing **isopanose** (O-α-D-glucopyra-nosyl-1,4-O-α-D-glucopyranosyl-1,6-D-glucopyranose). A third enzyme, *neopullulanase* from *Bacillus stearothermophilus*, has recently been des-cribed. The structural gene for the enzyme has also been cloned and expressed in *Bacillus subtilis*. The enzyme hydrolyses 1,4-α-D-glucosidic linkages in pullulan, yielding **panose** (O-α-D-glucopyranosyl-1,6-α-O-glucopyranosyl-1,4-D-glucopyranose). The mechanisms of action of these enzymes on their substrate are shown in Fig. 4.1. Pullulan can also be hydrolysed by glucoamylase (E.C. 3.2.1.3), the product from the non-reducing termini being glucose.

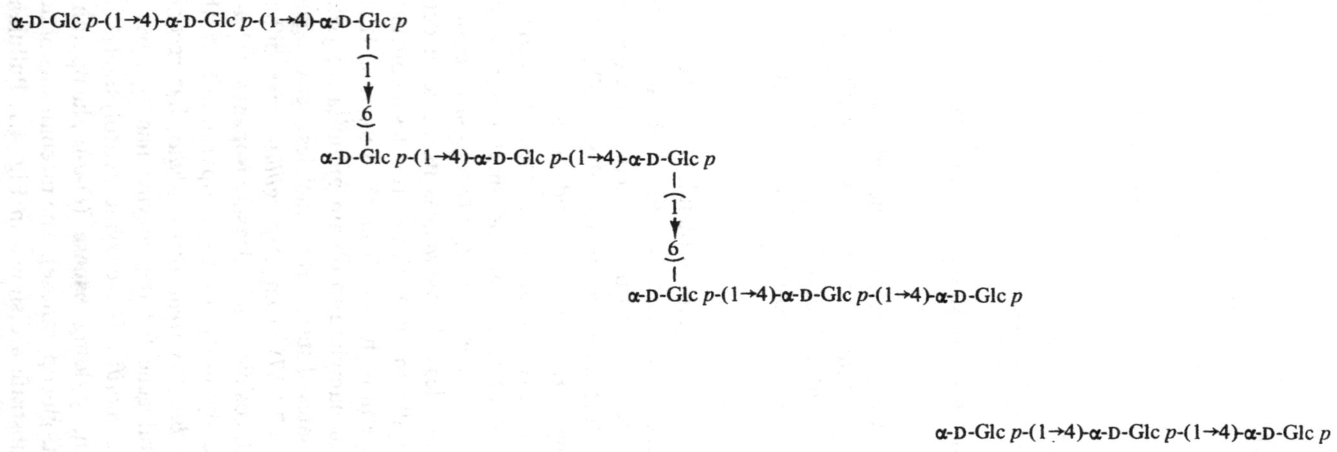

Fig. 4.2. Typical oligomeric products from the action of pullulanase on pullulan.

Several sources of pullulanase have been identified. Enzymes from *Klebsiella aerogenes* and from *Bacillus* species have similar modes of action; a thermostable and acidophilic pullulanase from another *Bacillus* strain is also available. The enzyme has proved to be of considerable value in structural analysis of branched chain α-D-glucans in addition to pullulan (Fig. 4.2).

As well as being susceptible to the enzymes indicated above, pullulan can be degraded by several other α-glucanases. Salivary α-amylase hydrolyses pullulan where there is a maltotetraosyl unit, while a similar enzyme from *Thermoactinomyces vulgaris* yields panose as the major product, together with small amounts of maltose, glucose and isomaltose.

The fungal product elsinan is degraded by some α-amylases, including one produced by *Aspergillus oryzae*. Some of these enzymes give high yields of 4-*O*-α-nigerosyl-D-glucose together with smaller amounts of D-glucose and the tetrasaccharide 3-*O*-α-maltosyl–maltose. The fungal enzyme yields a series of oligosaccharides with degrees of polymerisation (DP) 4, 7, 10, 13 and 16 with regularly arranged 1,3-α-D-glucoside linkages.

Xanthan-degrading enzymes

Various enzymes degrading xanthan are known; the lyases acting on the trisaccharide side-chain will be discussed separately. Although almost all xanthan-producing strains of *Xanthomonas* pathovars secrete cellulases, xanthan, despite its cellulosic main-chain structure, is not degraded by these. This is undoubtedly due to the ordered solution conformation adopted by xanthan at ambient temperatures in the presence of salts. When present in the *unordered* form, xanthan is susceptible to the action of a number of 1,4-endo-β-glucanases (cellulases). These enzymes then cause random cleavage of the polysaccharide backbone structure, leading to decreased solution viscosity and release of oligosaccharides (oligomers of the pentasaccharide repeat unit). Small quantities of glucose are also released. As it is extremely difficult to convert **all** the xanthan molecules to the unordered conformation at temperatures at which even thermostable cellulases will function, not all the polysaccharide is depolymerised. It is also possible that certain portions of the xanthan molecule are more susceptible to enzymic attack, perhaps owing to missing side-chains (Fig. 3.12). The reduction in molecular mass can be followed by gel permeation chromatography (Fig. 4.3). The products vary slightly depending on the xanthan used as substrate and on the endoglucanases, but thermostable enzyme preparations, from *Aspergillus niger*, *Trichoderma viride* and other fungi all show activity. The commercial enzyme preparations with such activity are all complex mixtures containing several β-D-glucosidases and endo-β-glucanases; exoglucanases may also be present.

Several enzyme systems hydrolysing xanthan in the ordered form have

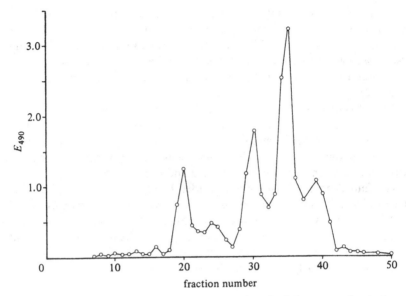

Fig. 4.3. The products obtained from the action of cellulase on xanthan in the unordered form, separated on a Biogel P6 column and detected as eluted carbohydrate in decreasing order of size.

also been described. One of these, containing an endo-β-glucanase active against xanthan, carboxymethylcellulose and some 1,3-β-D-glucans, caused rapid loss of viscosity and release of oligosaccharides. It was associated with lyase activity, from which it could be separated chromatographically. Both activities were induced following bacterial growth on xanthan as sole carbon substrate. A similar endoglucanase has been purified from a mixture of salt-tolerant bacteria. This enzyme has a pH optimum near 5.0 and a temperature optimum of 35 °C. The products from xanthan digestion were oligosaccharides ranging in size from 1 to 12 repeat units, together with polymer of residual relative molecular mass 8×10^5. This was interpreted as indicating that certain portions of the xanthan molecule were particularly susceptible to enzyme cleavage. The salt-tolerant endoglucanase has relatively narrow substrate specificity, only carboxymethylcellulose being readily hydrolysed, in addition to xanthan (Table 4.3).

Another enzyme preparation, also obtained from several mixed bacterial cultures, lacked endoglucanase but possessed various other activities. These cleaved all side-chain sugars, releasing D-glucose, D-mannose, O-acetyl-D-mannose, pyruvylated mannose and D-glucuronic acid. Yields of the enzymes depended on the component strains used in the mixed culture. The different enzymes identified as hydrolysing xanthan are shown in Fig. 4.4. It is clear that one of the problems in obtaining enzymes to study

Table 4.3. *Enzymes degrading xanthan*

Source of enzyme	Products
Bacillus sp.	endoglucanase, lyase
Bacillus sp.	endoglucanase, lyase, etc.
Bacillus spp.	endoglucanase, endomannosidases, endoglucuronidase
Bacillus mixed culture	endoglucosidase, etc.
Corynebacterium sp.	endoglucanase, lyase, etc.

Note:
All systems yielded complex mixtures of oliogosaccharides.

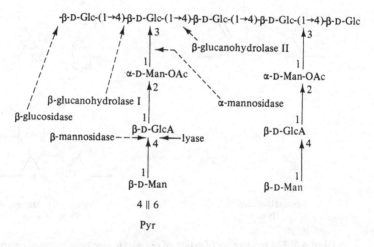

Fig. 4.4. The various sites of enzyme action identified on xanthan.

fragments from the xanthan molecule is that they have almost all been obtained from mixed bacterial cultures and few of the enzymes have been purified to homogeneity. This problem is of course a general one, and is not confined to enzymes acting on xanthan.

4.3.   Lyases (eliminases)

In contrast to hydrolytic enzymes, polysaccharide lyases cleave glycosidic linkages through an elimination mechanism. The products are unsaturated oligosaccharides into which, during cleavage of the main-chain bond adjacent to a uronic acid residue, a double bond is introduced into the acid sugar. This results in a product terminating in a 4-deoxy-5-oxo-uronic acid. The enzyme activity can be quantified either through measurement at

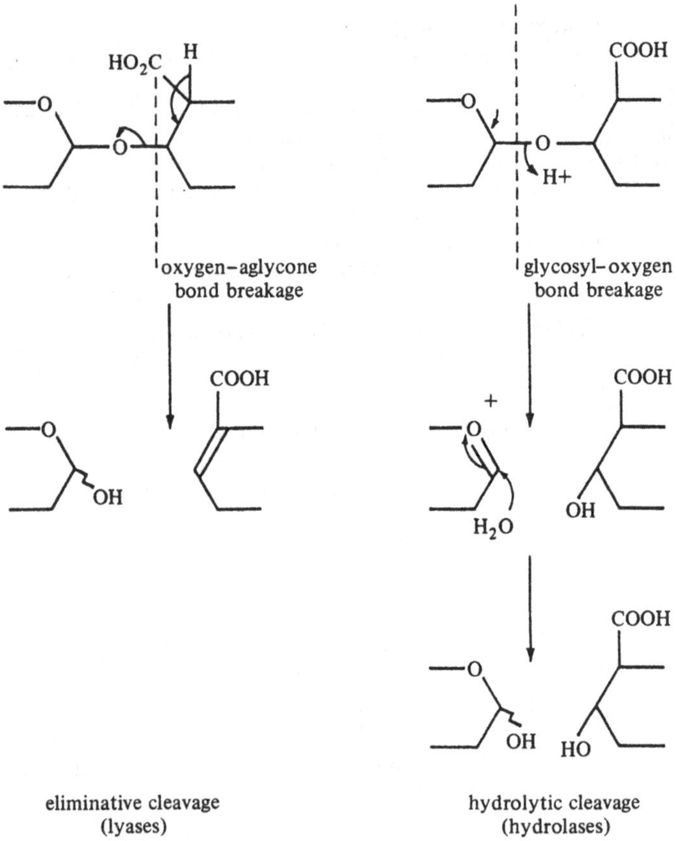

Fig. 4.5. The mechanism of action of polysaccharide lyase enzymes compared with that of hydrolases.

235 nm of the chromophore generated, or by interaction, after periodate oxidation, with thiobarbituric acid.

Lyases are widely distributed and their action as pectin-, heparin-, chondroitin- or hyaluronate-lyases against eukaryotic polysaccharides and proteoglycan substrates has been extensively studied. The mechanism of enzyme action can be compared to hydrolytic cleavage as shown in Fig. 4.5. The polysaccharide substrate contains a carboxylic acid group adjacent to the glycosidic linkage at the site of enzyme action. Abstraction of the proton through enzyme action causes a direct eliminative cleavage and produces an α,β-unsaturated uronic acid as the non-reducing terminal at the cleavage site. A hemiacetal is formed on the reducing side of the cleavage site. The majority of the lyases acting on microbial exopolysaccharides are **endo**enzymes, randomly cleaving the main chain of the

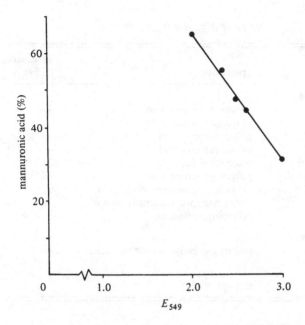

Fig. 4.6. Alginase action on substrates. Enzyme activity measured colorimetrically against alginates of differing mannuronic acid content.

polymers internally. However, one enzyme functioning as an **exoenzyme**, sequentially removing oligosaccharides from the polysaccharide chain, has recently been reported. The **xanthan lyases** (v.i.) differ from the other enzymes of this type in attacking the *side-chain* of the substrate polysaccharides.

Alginate lyases (E.C. 4.2.2.3)

Alginate prepared from either marine algae or bacterial exopolysaccharides is the substrate for a range of different *alginases*, all **endo**lyases. Although the enzymes may be specific to L-guluronic acid or D-mannuronic acid, the modification of the aglycone residue during enzyme action yields the same product from each of the original uronic acids. Consequently, alginases do not provide as much information on polysaccharide structure as do many other polysaccharases. They can nevertheless be used to produce fragments for study by NMR and other physical techniques. Most alginases show high specificity for **homopolymeric** substrates (poly-D-mannuronic acid or poly-L-guluronic acid) and their activity against different alginates is therefore a function of the substrate composition and of the relative proportion of the two component uronic acids (Fig. 4.6). It is also affected by the presence of the *O*-acetyl groups attached to D-mannuronic acid residues in bacterial alginates. Relatively little is known

Table 4.4. *Source and specificity of alginate lyases*

Source	Specificity	Products (DP[a])
Gram-negative bacteria		
Azotobacter vinelandii	polymannuronic acid	—[b]
A. vinelandii phage	polymannuronic acid	—
Beneckea pelagia	polyguluronic acid	—
Klebsiella pneumoniae	polyguluronic acid	—
Pseudomonas aeruginosa	polymannuronic acid	2, 2 +
P. maltophilia	polymannuronic acid	—
Marine bacterium A3	exolyase, polymannuronic acid	1
Marine bacterium	endolyase, polymannuronic acid	—
Marine bacterium	polymannuronic acid	3, 4, 4 +
Gram-positive bacteria		
Bacillus circulans	endolyase, polymannuronic acid	—
Eukaryotes		
Dendryphiella salina	not known	

Notes:
[a] DP, degree of polymerisation.
[b] Dashes indicate that the products have not been characterised.

about the effect of different heterologous uronic acid sequences in the substrate on enzyme action or that of the *O*-acetyl groups.

There is a wide range of sources from which alginases can be obtained. These include marine gastropods, prokaryotic and eukaryotic microorganisms and bacteriophage. Surprisingly, two of the bacterial species that produce alginates, *Pseudomonas aeruginosa* and *Azotobacter vinelandii*, also yield alginases. As the enzymes are normally located intracellulary (in the periplasm) in both microorganisms, they may not affect polysaccharide yields too greatly. The *P. aeruginosa* enzyme also has relatively low activity against the naturally acetylated alginate with high D-mannuronic acid content synthesised by these bacteria. The sources of some of the alginases which have been described are indicated in Table 4.4. All these enzymes are active in the degradation of algal alginate and many have been tested by using bacterial alginates as substrates. They do provide a means of studying the bacterial polysaccharides and may yield structural information to supplement that obtained by chemical and physical means.

Hyaluronidase (E.C. 4.2.99.1)

Hyaluronate lyases may be of animal or bacterial origin (hydrolytic enzymes of animal origin are also known). Most of the bacterial enzymes yield a disaccharide derived from the repeating unit structure of hyaluronic

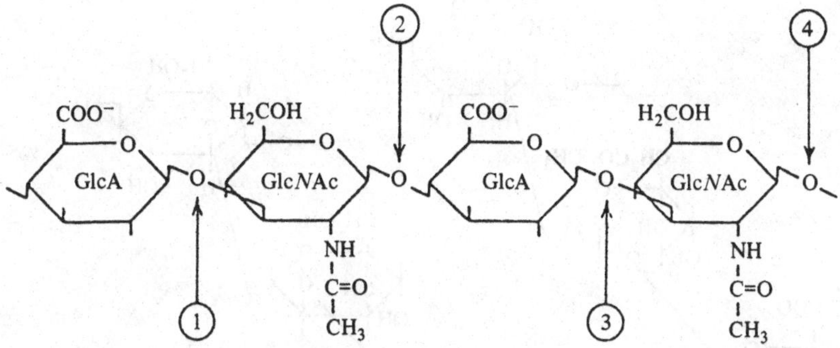

Fig. 4.7. The action of enzymes on hyaluronic acid. Numbers 1 and 3 show the sites of action of β-glucuronidases, and 2 and 4 those of β-*N*-acetylglucosaminidases.

acid, although larger fragments may occasionally be formed. One enzyme, from *Streptococcus pneumoniae*, resembles testicular hyaluronidase in its site of action and products (Fig. 4.7). Other enzymes have been obtained from the bacteria that synthesise hyaluronic acid, and from bacteriophage. The enzymes differ in their specificity: some also act on chondroitin and heparin, but that from a *Streptomyces* species only acts on hyaluronic acid. The products from the action of this enzyme are a tetrasaccharide and a hexasaccharide. These two oligosaccharides are not degraded further by this particular enzyme but can be used as substrates for other 'hyaluronidases'. An enzyme from a *Streptococcus* phage also produces larger fragments: a tetrasaccharide and an octasaccharide.

Xanthan lyase

As well as the hydrolytic enzymes acting on xanthan, which have already been described, xanthan lyases active on the trisaccharide side-chain of the polysaccharide have been found. The first such enzyme was obtained from a *Bacillus* species isolated from soil. Subsequently, similar enzymes were found in several bacterial species including another *Bacillus* isolate, a *Corynebacterium* species and a mixed bacterial culture (Sutherland, 1987). The enzymes were all associated initially with 'xanthanase' (1,4-**endo**-β-glucanase) activity, but the *Bacillus* enzymes could be separated by fast protein liquid chromotography. The lyase resembles several other such enzymes, being a β-mannolyase, but differed from the enzymes described below in cleaving a side-chain linkage rather than the main polysaccharide chain. The lyase is active against a range of different xanthans with and without acetyl and pyruvate ketal substituents on the side-chains. No activity was found against several other bacterial exopolysaccharides for which structural similarity to xanthan has been claimed. As the optimal size of the substrate appeared to be 5–7 repeat units of the polysaccharide, the

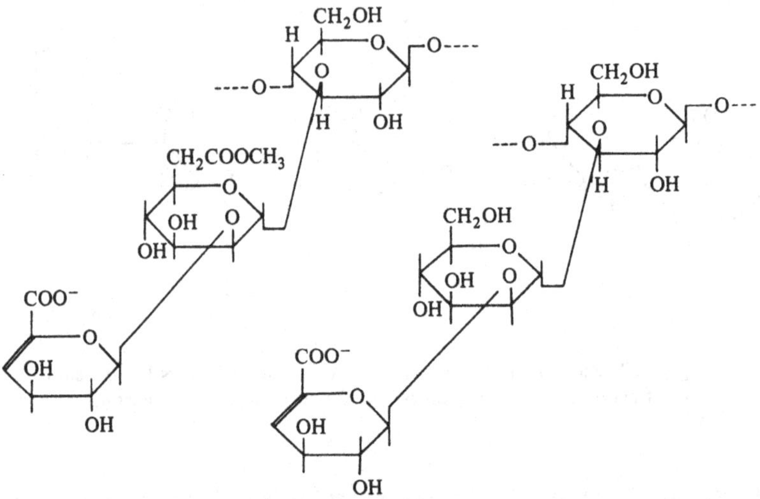

Fig. 4.8. Oligosaccharides obtained from xanthan through the combined action of lyase and endo-β-glucanase enzymes.

activity of the xanthanase may first be needed to reduce the molecular mass of the polymer. The combined action of the xanthanase and the lyase yields a series of oligosaccharides, each with a side-chain terminating in an unsaturated uronic acid and containing D-mannose and D-glucose in the molar ratio 1 : 2. The fragments generated by the two enzymes can provide useful information on aspects of the fine structure of xanthan (Fig. 4.8). The enzymes, and the lyase in particular, can also be used to provide the basis for specific assays, which can enable the quantitation of xanthan in foods and other industrial products. As the lyase is so specific in its action, its use permits the determination of xanthan even when other polysaccharides such as galactomannans or carboxymethylcellulose may be present in the formulation.

Emulsan depolymerase

The emulsifying polyanionic product **emulsan** from *Acinetobacter calcoaceticus* can be degraded by an enzyme from a bacterial isolate provisionally identified as a *Zoogloea* species. The enzyme, induced by its substrate, is a lyase, the action of which rapidly reduces both the viscosity and the emulsifying activity of emulsan. It is therefore useful in studying the role of the carbohydrate portion of the molecule in emulsifying action. Cleavage of less than 0.5% of the glycosidic linkages is sufficient to decrease emulsifying activity by more than 75%. The enzyme releases oligosaccharides of average relative molecular mass 3000 from the polysaccharide portion of the emulsan molecule. Analysis of the residual material produced by enzyme action showed a relatively constant ester content, indicating that the esters were not uniformly distributed on the polymer molecule. Such results can only be obtained through analysis of enzyme-generated fragments and indicates the value of enzymes as analytical tools for this purpose.

Other polysaccharide lyases

Apart from lyases active against microbial alginate and against hyaluronic acid, several other such enzymes have been recognised. Most have been obtained from bacteriophages. Two are **endo**mannolyases, although they differ in their specificity for the polysaccharides of *Klebsiella aerogenes* serotype K5 and K64 respectively. The main-chain sequence in each polymer is:

β-D-Glc$p$→D-Man$p$→D-Glc$p$A . . .

However, in the K5 exopolysaccharide, the D-glucosyl bond is 1,3; in K64 the linkages are 1,4 and α- respectively. The xanthan lyase already mentioned is also a mannolyase but fails to act on either of the *Klebsiella aerogenes* polysaccharides. The two phage mannolyases are **endo**enzymes, causing rapid loss in the viscosity of the substrate polysaccharide solutions. As the xanthan lyase attacks side-chains, it is effectively an **exo**mannolyase. A further **exo**lyase of this type has been found which acts on **gellan**. The specificity of this enzyme is different from that of the other lyases discussed in this section. The substrate contains a D-glucosyl-D-glucuronosyl linkage, which is the site of enzyme cleavage. The enzyme releases a trisaccharide product, although the repeating unit of the exopolysaccharide is a tetrasaccharide. It may thus require the presence of an associated β-D-glucosidase or endo-β-glucanase to remove the fourth sugar. There is no activity against other polysaccharides of the gellan series, the structures of which are shown in Fig. 3.7, probably owing to steric hindrance by the side-chains or the postulated irregularity of some of the main-chain structures.

Further reading

Aspinall, G. O. (1985). *The polysaccharides*, vol. 3. Academic Press, Orlando.

Berkeley, R. C. W., Gooday, G. W. & Ellwood, D. C. (1979). *Microbial polysaccharides and polysaccharases*. Academic Press, London.

Brown, R. D. & Jurasek, L. (1979). Hydrolysis of Cellulose: Mechanisms of enzymatic and acid hydrolysis. *ACS Symposium 181*. American Chemical Society, Washington.

Rieger-Hug, D. & Stirm, S. (1981). *Comparative Study of Host Capsule Depolymerases associated with* Klebsiella *Bacteriophages. Virology* **113**, 363–78.

Tipson, R. S. & Horton, D. (eds.) (1986). *Advances in Carbohydrate Chemistry and Biochemistry*, vol. **44**. Academic Press, London.

5 Biosynthesis

5.1. Introduction

The biosynthesis of most exopolysaccharides closely resembles the process by which the bacterial wall polymers peptidoglycan and lipopolysaccharide are formed. Indeed, the three types of macromolecule share the characteristic of being formed of carbohydrates and associated monomers, being synthesised at the cell membrane and exported to final sites external to the cytoplasmic membrane. The only exceptions are the exopolysaccharide levans and dextrans, which are synthesised by a totally extracellular process and whose formation will be discussed later in this chapter.

Formation of the precursors for polysaccharide synthesis occurs within the cytoplasm. This is probably a necessity to ensure that they are thus readily available, as in many cases they are utilised for several different polymer-synthesising systems. As they are freely soluble in the cytoplasm, they can be readily channelled to the appropriate biosynthetic process occurring at or within the cytoplasmic membrane. Elucidation of the initial stages of polysaccharide synthesis has proved more difficult than was the case for polymers found in microbial and particularly bacterial walls. This has been mainly because of the lack of suitable selection systems for obtaining mutants and of antimicrobial agents specifically inhibiting polysaccharide biosynthesis. Even the preparation of cell-free systems or of membrane fragments is rendered more difficult by the presence of the viscous extracellular polysaccharides. Various cell-free systems, including membrane fragments from ultrasonically lysed cells, solvent-extracted cells or cells permeabilised with solvents or chelating agents, have been used.

Mutants have proved useful in studies where it has been possible to obtain microorganisms deficient in precursor synthesis (UDP-glucose pyrophosphorylase or UDP–galactose-4-epimerase, etc). These have the advantage of being low or lacking in endogenous sugar nucleotides and can be used to study the formation of intermediates and polymer from sugar nucleotides appropriately labelled in the sugar moiety. A few mutants known to be defective in later stages of polymer synthesis have also been studied. Mutants have also been essential for genetic studies (Chapter 7).

5.2. Carbohydrate precursors

For those exopolysaccharides that are synthesised at the cell membrane, there is a requirement for activated precursors: energy-rich forms of the

54

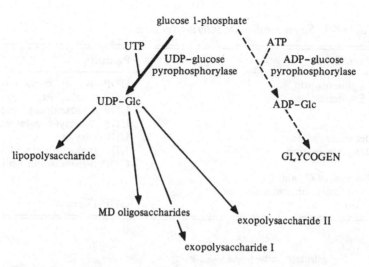

Fig. 5.1. Routes of anabolism of glucose in Gram-negative bacteria.

monosaccharides. These are mainly **nucleoside *di*phosphate** sugars, but a small number of nucleoside *mono*phosphate sugars also function in this role. These **sugar nucleotides** function in the synthesis of virtually all carbohydrate-containing polymers found outside the microbial cell membrane. Thus, in bacteria, they are also needed for the synthesis of wall polymers and, in many species, for the production of **membrane-derived oligosaccharides** (MDO) found mainly in the periplasm of Gram-negative species. Sugar nucleotides are also required for the synthesis of such intracellular storage products as glycogen and (mainly in eukaryotes) the disaccharide trehalose. The possible fate of a carbohydrate substrate in terms of catabolism and anabolism is shown in Fig. 5.1. The manner in which the carbohydrate substrate and the sugar nucleotides are channelled into the production of these different molecules is still unclear, but represents one aspect that is critically dependent on the physiological conditions employed for microbial growth. Only one example is known so far in which a distinct sugar nucleotide is used for the formation of one specific glycan. Thus in prokaryotes, ADP–D-glucose is used in glycogen synthesis, whereas *all* other glucose-containing polymers appear to be formed from UDP–D-glucose. In this way, the synthesis of the storage polymer glycogen can be regulated in a manner distinct from glycan formation. Although the allosteric regulators for ADP–glucose pyrophosphorylase activity differ in the various bacterial species, they are all primary products of carbohydrate catabolism or of gluconeogenesis.

The sugar nucleotides serve various functions: they are activated monosaccharide intermediates (capable of releasing 31.8 kJ mol^{-1} on hydrolysis, compared with the 20 kJ mol^{-1} from a sugar phosphate)

Table 5.1. *Sugar nucleotide interconversions*

Mechanism	Products
2-Epimerisation	UDP–N-acetyl-D-mannosamine
4-Epimerisation	UDP–D-galactose,
	UDP–D-galacturonic acid,
	UDP–N-acetyl-D-galactosamine
Decarboxylation	UDP–D-xylose
Dehydrogenation	UDP–D-glucuronic acid,
	GDP–D-mannuronic acid
Reversal of C_3 and C_5 configuration and deoxysugar formation	GDP–L-fucose

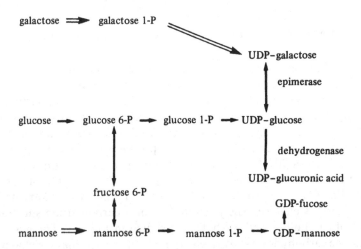

Fig. 5.2. Catabolic systems leading to sugar nucleotide synthesis and interconversions. Open arrows indicate catabolic systems used on specific substrates.

supplying the energy needed for assembly of oligosaccharide sequences on appropriate carrier molecules. They also provide a means of interconversion of different monosaccharides. This can be achieved through various mechanisms: epimerisation, oxidation, decarboxylation, reduction and rearrangement. Examples of these different mechanisms and their products can be seen in Table 5.1; interconversion of some of the more commonly utilised sugar nucleotides is indicated in Fig. 5.2. With the discovery that D-galacto**furanose** is a component of some exopolysaccharides, it became clear that another role of the sugar nucleotides was both that of providing a mechanism of interconversion of the pyranose to the furanose form, and that of yielding an activated form of the latter. This is achieved through the normal galactose-containing intermediate:

UDP–D-galacto**pyranose**→UDP–D-galacto**furanose**.

However, evidence from the algal species *Chlorella pyrenoidosa* indicates that L-galactose, which is also found in a small number of polysaccharides, is formed by a reversible epimerisation from GDP–D-mannose, the proposed mechanism being that indicated in Fig. 5.3. There may well be other such pathways to provide unusual forms of sugars.

Some of the sugar nucleotides appear to function only as intermediates in the synthesis of other sugars. TDP–D-glucose or CDP–D-glucose are not known to provide glucose in an activated form for polysaccharide synthesis, but each is an important intermediate: in the synthesis of TDP–rhamnose and of CDP–3,6-dideoxyhexoses (abequose, colitose, paratose and tyvelose) respectively. The majority of sugar nucleotides can be seen to be nucleoside **di**phosphate sugars, but one important nucleoside **mono**phosphate sugar is involved in exopolysaccharide synthesis: CMP–sialic acid. Another CMP–nucleotide, CMP–ketodeoxyoctonic acid, plays an important role in the biosynthesis of cell-wall lipopolysaccharides in Gram-negative bacteria. It is presumably also the precursor for KDO-containing exopolysaccharides, including those which have recently been found in several *Escherichia coli* serotypes.

Almost without exception, the discovery of a monosaccharide in an extracellular polysaccharide implies that a sugar nucleotide will be required to provide an activated form for transfer to the polymer. Loss of the capacity for synthesising the sugar nucleotide usually leads to loss of polysaccharide production. A few exceptions to this exist. As can be seen from Fig. 5.2, if certain mutants are supplied with the appropriate monosaccharide, it **may** be possible, through phosphorylation of the sugar and subsequent involvement of a *nucleotide sugar pyrophosphorylase*, to circumvent the part of the biosynthetic pathway lost by mutation. Thus, growth of UDP–galactose-4-epimerase-less mutants on galactose *may* overcome the loss of the enzyme through the phosphorylation of galactose by *hexokinase* or *galactokinase and* its activation as UDP–galactose.

In addition to the well-characterised series of interconversions of sugars at the sugar nucleotide level shown in Fig. 5.2, other reactions yielding less commonly found sugars have been studied. Extracts from cells of *Streptococcus pneumoniae* type XIV proved capable of forming a **di**amino sugar nucleotide:

UDP–GlcNAc→UDP–2-acetamido-4-keto-2,6-dideoxyhexose
$\qquad\qquad\qquad\qquad$↙——pyridoxal phosphate
$\qquad\qquad\qquad\qquad$↓$\qquad\qquad$+ glutamate
α-oxoglutarate + UDP–2-acetamido-4-amino-2,4,6-trideoxyhexose.

Micrococcus luteus is the source of polysaccharides containing two

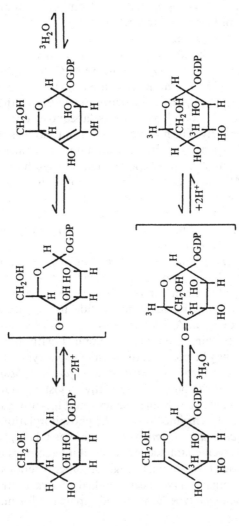

Fig. 5.3. The formation of GDP–L-galactose from GDP–D-mannose.

aminouronic acids, both of which are formed from the corresponding UDP–amino sugars by specific dehydrogenases:

UDP–N-acetylglucosamine→UDP–N-acetylglusosaminuronic acid; and

UDP–N-acetylmannosamine→UDP–N-acetylmannosaminuronic acid.

No evidence has been found for epimerisation at the aminouronic acid level. However, in both *Escherichia coli* and *Achromobacter georgipolitanum* cells, each of which produces mannosaminuronic acid-containing exopolysaccharides, two enzymes have been identified. The amino sugar moiety is epimerised at the C_2 position to yield UDP–N-acetylmannosamine, which is then oxidised at the C_6 position by an NAD-linked dehydrogenase.

Several types of microbial cells have the capacity to produce sugar nucleotides for which no function has yet become apparent. Some of these compounds *may* be involved in polysaccharide synthesis and *may* provide further examples of preferential use of a specific sugar nucleotide for one particular polymer.

5.3.   Acyl precursors

As well as activated forms of the monosaccharide components, the synthesis of exopolysaccharides demands precursors of any other groups present in the biopolymer. Thus the biosynthetic mechanism needs to have available to it activated forms of acetate, pyruvate, succinate, 3-hydroxybutanoate, phosphate and (in cyanobacterial systems) sulphate. Studies in other polymer systems clearly indicate that acetyl groups are added from acetyl CoA and this has been confirmed for the synthesis of the acetyl groups on the internal D-mannosyl residues of xanthan from *Xanthomonas campestris* pv. *campestris* as well as for some other bacterial exopolysaccharides. A similar result has been obtained when permeabilised cells were used to synthesise the acetylated octasaccharide repeat units of *Rhizobium trifolii* polysaccharide.

Although no studies on cell-free systems from bacteria producing polysaccharides with more exotic acyl groups have yet been reported, it seems likely that they, too, are derived from the corresponding acyl CoA derivatives. Such acyl compounds are needed for many other processes within the microbial cells and are always likely to be present. Their use for the 'decoration' of exopolysaccharides may be strictly regulated and dependent on the intracellular concentration. If the amounts of intracellular acyl derivatives are low, the priority of the polysaccharide may also be low, yielding either non-acylated polymer or material with greatly reduced acyl content. This might account for the results obtained for xanthan and other polysaccharides produced at certain stages of the bacterial growth cycle or under certain physiological conditions of nutrient limitation.

5.4.   Methyl groups

O-methyl sugars are present in polysaccharides from prokaryotes as well as algae, fungi and some lower eukaryotes. Therefore, although not uncommon, they are usually relatively minor components of the polymers in which they are found. The biosynthesis of the *O*-methyl sugars in exopolysaccharides has not been studied, but in both eukaryotes and prokaryotes, wall polysaccharides in which these modified sugars are found derive the methyl groups from methionine or *S*-adenosylmethionine. Studies of 3-*O*-methylation of D-mannose residues in *Mucor rouxii* (Lederkrener & Parodi, 1984) indicates that methyl groups are transferred **after** the monosaccharides have been released from isoprenoid lipid (dolichol) and linked to glycoprotein. In *Mycobacterium smegmatis*, biosynthesis of an α-D-mannose-containing polysaccharide utilises *S*-adenosylmethionine (Kamisango et al., 1987). The polymer is elongated by a sequential mannosylation–methylation reaction and chains are terminated by non-methylated sugar residues. Similarly, an *O*-methylglucose-containing lipopolysaccharide is probably formed by an alternate glucosylation–methylation sequence. In *Dictyostelium discoideum*, glycoproteins containing *O*-methyl sugars derive the methyl group from methylmethionine *in vivo* and from methyl-*S*-adenosylmethionine *in vitro* (Freeze & Wolgast, 1986). It is likely that *O*-methyl sugars are introduced into exopolysaccharides by similar mechanisms and it probably occurs at the isoprenoid-lipid-linked oligosaccharide level in a manner analogous to acetylation and ketalation.

5.5.   Amino acids

It is not yet known how amino acids are added to exopolysaccharides. Possibly a mechanism similar to that involved in peptidoglycan synthesis is needed. In this, ATP and a specific transferase with Mn^{2+} as cofactor, are involved in addition of amino acids at the sugar nucleotide level. Again, as with other substituents, addition of the amino acids to exopolysaccharides would probably be to lipid-linked oligosaccharide intermediates.

5.6.   Inorganic substituents

Phosphate groups may be derived from ATP by an appropriate phosphorylation reaction, but in many cases are more likely to emanate from the sugar nucleotide. Thus ribitol- and glycerol-containing teichoic acid-like polymers are formed from CDP–ribitol and CDP–glycerol respectively:

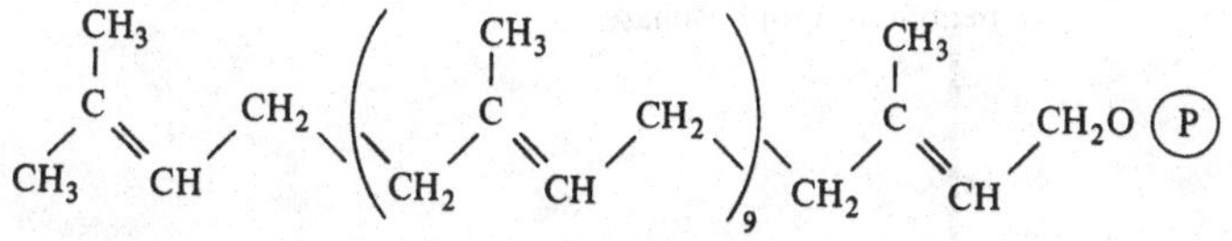

Fig. 5.4. The structure of the metabolically active form of bactoprenyl phosphate.

CDP–D-ribitol→(ribitol.P . . .) + CMP;
CDP–D-glycerol→(glycerol.P . . .) + CMP.

Neither the source of sulphate groups in microbial polysaccharides, nor the mechanisms of their addition, have been examined. The activated donor for polysaccharide sulphation in eukaryotic systems such as heparin, chondroitin and dermatan, is 3′-phosphoadenylyl sulphate (PAPS). For these polymers, specific *N*-sulphotransferases and *O*-sulphotransferases are involved. In the biosynthesis of the sulphated cyanobacterial polysaccharides, the sulphation mechanism is probably similar.

5.7. Lipid intermediates

The requirement for a lipid acceptor, on which to assemble the repeating units of the polysaccharides, is common to all such polymers found external to the microbial cell membrane. The discovery of the involvement of **isoprenoid lipids** (bactoprenol) in the formation of lipopolysaccharides and of peptidoglycan, indicated that they might also have a role in exopolysaccharide synthesis. Studies with a range of *Klebsiella aerogenes* mutants has shown the sequential addition of D-glucose-1-phosphate and two moles of D-galactose from UDP–D-glucose and UDP–D-galactose, respectively, to form lipid-soluble oligosaccharides. In a further study with another strain of the same species, the lipid has been extracted and characterised. The functional lipid in prokaryotic polysaccharide synthesis is a C_{55} isoprenyl phosphate (syn. bactoprenyl phosphate, undecaprenyl phosphate) (Fig. 5.4). This is the same acceptor lipid which functions in the formation of peptidoglycan and the side-chains (*O*-antigens) of lipopolysaccharides. In bacteria, the total isoprenoid lipid available in each cell has been estimated as 5.5×10^4 molecules, equivalent to 0.02% of the cell's dry mass. This lipid, because of its function in several polysaccharide-synthesising systems, must be assumed to be under some type of control in terms of its availability. When extracted from bacterial cells, it is found in the form of the free alcohol, alcohol phosphate and alcohol pyrophosphate. Enzymes effecting the interconversion of these forms have been reported from Gram-positive bacteria and are indicated in Fig. 5.5. The phosphorylation and dephosphorylation of the bactoprenyl derivatives **may** provide a means of

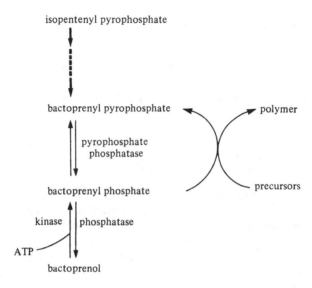

Fig. 5.5. The activation and deactivation of isoprenoid lipid. The polymer may be exopolysaccharide, lipopolysaccharide or peptidoglycan.

controlling their availability but it is unclear whether the necessary enzymes are present in all prokaryotic cells. Even less is known about eukaryotes, in which a similar lipid of longer chain length, dolichol ($C_{70}-C_{90}$) appears to perform a similar function for some polysaccharide-synthesising systems. However, it should not be assumed that the process of exopolysaccharide synthesis is exactly the same as that in prokaryotes.

A series of studies on *X. campestris* has confirmed that the synthesis of xanthan followed the same pattern as has been found in *Klebsiella* and other bacterial species. However, the study was extended to include acyl groups. Through the use of isotopically labelled sugar nucleotides and acyl precursors, it has confirmed that acetate and pyruvate, like the monosaccharides, are added sequentially to the lipid carrier. Thus D-glucose-1-phosphate is added initially from UDP–D-glucose. Thereafter, D-glucose, D-mannose and acetate are transferred. Finally, D-glucuronic acid, D-mannose and pyruvate (as a ketal) are added fromn UDP–D-glucuronic acid, GDP–D-mannose and phosphoenolpyruvate respectively, thus forming the acylated pentasaccharide repeat unit (Fig. 5.6).

Following assembly of the repeat unit on the isoprenoid lipid, polymerisation occurs to yield acetylated and pyruvylated xanthan. The initial pentasaccharide is essentially linear and branching only occurs as the oligosaccharide repeat units are polymerised. As judged by studies on other polysaccharide-synthesising systems, polymerisation occurs by addition to the non-reducing terminus, while attached to isoprenoid lipid pyrophos-

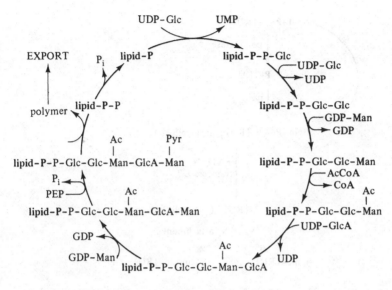

Fig. 5.6. A scheme for xanthan biosynthesis (after Ielpi *et al.*, 1981, 1983.)

phate. The larger fragment is transferred to lipid carrying a single repeat unit. This process continues until the macromolecule is formed.

The polymerase appears to have relatively low specificity in the case of *X. campestris*. A number of studies have suggested that the polysaccharide molecules are imperfect: lacking side-chains or carrying incomplete side-chains. A mutant has been obtained which is capable of the polymerisation of the cellobiosyl intermediate to form a cellulosic polysaccharide and of the mannosylated cellobiose (linked to lipid) to yield a 'polytrisaccharide'. This latter polymer can be synthesised in fairly good yield and, as judged by its viscosity, the product has a molecular mass comparable to wild type xanthan. Whether there is a specific mechanism for the regulation of the molecular mass of an exopolysaccharide is unclear (see p. 97). The molecular mass of the product probably depends both on the strain and on the physiological conditions under which the microorganisms are grown. In batch culture at least, the molecules are probably polydisperse; the range of molecular mass obtained may depend to a large extent on the microbial species under study. It is by no means clear how the molecular mass is controlled. Stable mutants with polysaccharides of increased viscosity (and molecular mass) have been obtained from several different Gram-negative bacterial species. In eukaryotic cells synthesising heparin, it has been suggested that the molecular mass of the polysaccharide depends on the relative amounts of the two precursor sugar nucleotides. There is obviously a considerable difference from lipopolysaccharide synthesis in Gram-negative

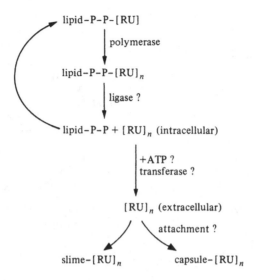

Fig. 5.7. The postulated final stages of exopolysaccharide synthesis and excretion. RU indicates a repeat unit, such as xanthan pentasaccharide.

bacteria, in which a wide range of molecular masses is found for the antigenic portion of these macromolecules, ranging from one repeat unit to thirty or more.

There is still relatively little known about the final stages of polysaccharide synthesis (Fig. 5.7). The site of assembly and polymerisation of the repeat units is almost certainly the cytoplasmic membrane. As the chain is elongated, the intermediate becomes increasingly hydrophilic, yet the reactions occur in a lipoprotein bilayer membrane. The exopolysaccharide must be released from the lipid pyrophosphate, passed to and extruded through the outer membrane (in Gram-negative bacteria) and released into the extracellular environment. Alternatively, capsular material must be attached to a specific component on the cell surface. Presumably, energy, in the form of proton-motive force and/or energy-rich phosphate bonds, is needed to permit the extrusion and release of the polymer, but the mechanism has not yet been elucidated. Comparison with lipopolysaccharide synthesis certainly indicates that a mechanism exists for supplying the energy necessary for translocating LPS from its site of synthesis in the cytoplasmic membrane to the outer face of the outer membrane. Exopolysaccharide formed by some mutants defective in transport may be found in the periplasm; in other (lethal) mutants it accumulates in the cytoplasm. It is possible that transfer of exopolysaccharide from the site of synthesis at the cytoplasmic face of the inner membrane is followed by chain termination at the periplasmic face, and transfer to the outer membrane may involve the fusion zones demonstrated by Bayer. These transient bridges between inner and outer membranes certainly appear to function in the export of the wall polysaccharides peptidoglycan and lipopolysaccharide from their site of synthesis in the cytoplasmic membrane to their final location in the periplasmic region and the outer membrane respectively. A recent study on sialic acid secretion has provided further genetic evidence for two distinct stages: transport of the exopolysaccharide from the site of synthesis to the periplasm, then further transport to the bacterial cell surface (Frosch *et al.*, 1989). No thorough investigation of the synthetic process in Gram-positive bacteria or in eukaryotes has yet been reported.

In one bacterial system, the ability to produce exopolysaccharide is apparently associated with the presence of a specific outer membrane protein. It is thought that this protein may provide a porin (water-filled channel) through which the polysaccharide could be exported. The diameter of the porin, 1.2 nm, is probably not great enough to allow ready egress of the macromolecule. It remains to be seen whether there is indeed a role for either specific or non-specific porins in the polysaccharide export process, although they do provide transmembrane proteins for the uptake of small molecules.

5.8.　Energy requirements

Few detailed studies have attempted to quantify the energy needed for the biosynthesis and excretion of exopolysaccharides. At high specific rates of synthesis, found when bacteria are grown in continuous culture, the demand for energy (in the form of ATP) represents a significant proportion of that of the cell. The synthesis of the polysaccharide represents a large portion of the biosynthetic activity of the microbial cell, under such conditions. Examination of the scheme proposed for xanthan biosynthesis (Fig. 5.6) indicates a requirement of 1 mole of ATP for each hexose substrate molecule converted to hexose phosphate. A further high-energy phosphate bond is needed for the synthesis of each sugar nucleotide, and further energy is required for phosphorylation of the isoprenoid lipid intermediates and for the acyl substituents. Presumably, polymerisation and extrusion of the polysaccharide also need energy. Jarman and Pace (1984) estimated that 11 moles of ATP were needed for the production of each pentasaccharide repeat unit of xanthan, whereas alginate was calculated to need 3 moles of ATP for *each* mannuronic acid or guluronic acid residue in the polysaccharide.

5.9.　Post-polymerisation modification

While a number of exopolysaccharides may be degraded after release from the microbial cell, by enzymes present in the extracellular environment, one group of polymers undergoes modification without reduction in molecular

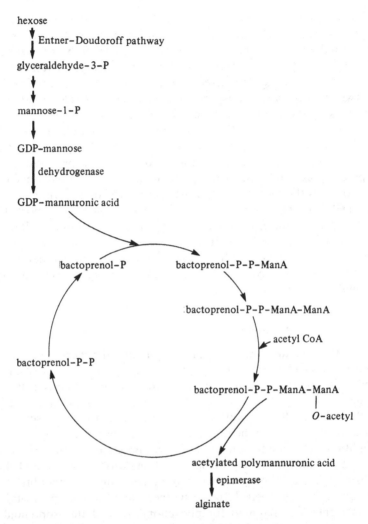

hexose

Entner–Doudoroff pathway

glyceraldehyde–3–P

mannose–1–P

GDP–mannose

dehydrogenase

GDP–mannuronic acid

bactoprenol–P   bactoprenol–P–P–ManA

bactoprenol–P–P–ManA–ManA

acetyl CoA

bactoprenol–P–P

bactoprenol–P–P–ManA–ManA

O–acetyl

acetylated polymannuronic acid

epimerase

alginate

Fig. 5.8. The probable mechanism of bacterial alginate synthesis.

mass. This type of **post-polymerisation epimerisation** is a feature of several polysaccharides of irregular structure found as cellular components (frequently as glycoproteins) in eukaryotes. Thus, in the synthesis of dermatan, C_5 inversion of D-glucuronosyl residues to L-iduronosyl residues is catalysed by a dermatan epimerase. A similar process is found in heparin biosynthesis, the epimerisation again involving conversion of D-glucuronic acid to L-iduronic acid in the polymer sequence. It is, however, unique to one group of bacterial polysaccharides: bacterial alginates. These lack a repeat unit and are highly charged polymers. The charge : mass ratio is higher than in any other microbial polysaccharides, as they are composed solely of D-mannuronic acid and L-guluronic acid. The alginates vary widely in the ratio of D-mannuronic acid to L-guluronic acid and in their acetyl content (Table 3.2). In a few preparations, the guluronic acid content is very low or apparently absent. In the absence of evidence to the contrary, it appears that the alginates are synthesised as **homopolymers** by mechanisms analogous to those found for such polysaccharides as xanthan, i.e. the mannuronic acid residues are thought to be assembled and acetylated on isoprenoid lipids and finally polymerised and excreted from the bacterial cell as acetylated poly-D-mannuronic acid. Subsequently, they are modified by an extracellular epimerase without the need of soluble cofactors (Fig. 5.8). The epimerase does require calcium ions for its activity and effects the exchange of H_5 of the mannuronosyl residue with water. C_5 inversion of the D-mannuronic acid residue to L-guluronic acid in alginates, and of D-glucuronic acid to L-iduronic acid in eukaryotic polysaccharides, occurs at the polymer level. The reactions show some similarity in that the epimerisation first requires abstraction of the hydrogen$_5$ of the target uronosyl residue. During inversion, the configuration at C_5 changes and a proton is exchanged with one from the aqueous environment. The role of the calcium ions in alginate epimerisation appears to be twofold. The reaction is stimulated and the concentration of calcium influences the epimerisation pattern. At low concentrations (*ca.* 0.9 mM), the reaction pattern of the epimerase from *A. vinelandii* favours introduction of neighbouring L-guluronosyl residues and promotes the creation of longer sequences of this uronic acid (i.e. poly-L-guluronic acid 'blocks'). In the presence of higher concentrations of calcium ions, the epimerase tends to introduce single guluronosyl units to form alginates with a considerable content of alternating structural sequences. There appear to be differences in the specificity of the epimerase enzymes from different alginate-producing bacterial species. As can be seen in Fig. 3.5, only *A. vinelandii* produces alginate with contiguous L-guluronic acid sequences. The alginate from the other bacterial species contains only single residues of this sugar. It has therefore been suggested that some other biosynthetic mechanism is involved in *Pseudomonas* alginate production, but no evidence for this has been presented.

The action of the epimerase is greatly affected by the presence of O-acetyl groups on the poly-D-mannuronic acid. These substituents exert a 'protective' effect on the monosaccharide residues, which inhibits their epimerisation (Fig. 5.9). The extent of epimerisation of bacterial alginates is thus dependent on several different factors, including the degree of acetylation, calcium ion concentration, and the bacterial species involved. Interestingly, the O-sulphation of either the L-iduronosyl residues (at C_2) or the vicinal D-glucosaminyl residues (at C_6) of heparin intermediates, also prevents 5-epimerisation and can be similarly considered to exert a protective effect.

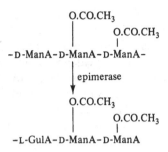

Fig. 5.9. Protection of D-mannuronic acid residues by acetylation.

5.10.   Extracellular synthesis

Only one polymer of industrial importance is synthesised extracellularly. Dextran is formed by a number of bacterial species and is produced commercially using strains of *Leuconostoc mesenteroides*. Unlike the exopolysaccharides that are formed from intracellular precursors and are synthesised from a wide range of growth substrates, dextran synthesis only occurs when the bacteria are grown on sucrose as carbon substrate. The polysaccharide can be produced either using whole bacterial cultures or cell-free preparations. While the latter have value for experimental studies, commercial processes utilise whole microorganisms in culture. The preparations must contain the enzyme *dextransucrase* (E.C. 2.4.1.5) (1,6-α-glucan:D-glucose 2-glucosyltransferase), which is induced in the presence of the substrate sucrose. Synthesis of the glucan requires cleavage of the substrate molecule to release free fructose, while the glucosyl residue is transferred to an appropriate acceptor, the actual mechanism involving transfer of the glucose to the reducing end of a nascent dextran chain. The acceptor is bound to the enzyme and the reaction catalysed is

$(1,6\text{-}\alpha\text{-}D\text{-glucose})_n$ + sucrose→$(1,6\text{-}\alpha\text{-}D\text{-glucose})_{n+1}$ + fructose.

The energy needed for synthesis comes in this reaction, not from a sugar nucleotide as has been noted for other polysaccharides, but from hydrolysis of the substrate sucrose. The free energy obtained from the hydrolysis of the sucrose molecule is 23 kJ mol^{-1}, enabling the enzymic reaction to proceed from left to right, as the glucosidic bond in dextran has a lower free energy level (12–17 kJ mol^{-1}). The dextran produced has a high molecular mass; no evidence has been found for any low-molecular-mass intermediates. Various enzyme preparations have been used, varying considerably in their purity; immobilised enzymes have also provided some information. The polysaccharide product may be composed of up to 95% 1,6-α-linked glucosyl residues, but 1,2-,1,3- and 1,4-α-linkages may also be present. Various mechanisms to account for the branching of the dextran molecule have been proposed. One suggestion is that a C(3)-OH on the acceptor molecule acts as a nucleophile on C(1) of the reducing terminal of a dextran–dextransucrase molecular complex. The polysaccharide is thus displaced from the enzyme and a 1,3-α-glucosyl branch is formed. The mechanism of synthesis differs from that involved in the intracellular α-D-glucan storage polymer, glycogen. There is no requirement for either a sugar nucleotide or a separate branching enzyme. Exogenous dextran generally stimulates synthesis, even when the dextran is chemically modified at the non-reducing terminus. This stimulation *may* result from displacement of the growing dextran chain from the active site of the enzyme, thus providing a nascent enzyme capable of reacting more rapidly with the substrate.

The dextransucrase of *Streptococcus mutans*, responsible for the synthesis of dextrans in which 1,3-α-linkages predominate, is constitutive, synthesis being growth-associated. Both cell-bound and soluble forms of the enzyme are found in cultures. The dextran product is also in two forms: soluble and insoluble.

The synthesis of most microbial exopolysaccharides shows many similarities to the production of other bacterial polysaccharides formed from repeating unit structures, i.e. lipopolysaccharide and peptidoglycan. Many of the precursors are sugar nucleotides common to these other polymers, and the same isoprenoid carrier lipids are used for assembly of the repeat units of each type of polymer. Export is probably an energy-requiring process but is, as yet, the least understood part of polysaccharide biosynthesis.

Further reading
Sharon, N. (1975). *Complex carbohydrates.* Addison-Wesley, London.
Stoddart, R. W. (1984). *The biosynthesis of polysaccharides.* Croom Helm, London.
Tonn, S. J. & Gander, J. E. (1979). Biosynthesis of polysaccharides by prokaryotes. *Annual Reviews of Microbiology* **33**, 169–99.
Troy, F. A. (1979). The chemistry and biosynthesis of selected bacterial capsular polymers. *Annual Reviews of Microbiology* **33**, 519–60.

6   Physiology and industrial production

6.1.   Introduction

In the research laboratory, the production of polysaccharides by either batch or continuous fermentation generally presents few problems. When, however, the process is scaled up and converted to an economic method for production, considerable difficulties may be encountered which are not apparent on the smaller scale. These include in particular oxygen transfer, heat transfer, substrate conversion and product recovery.

6.2.   Influence of medium composition

The majority of polysaccharide-producing microorganisms that have been studied utilise carbohydrates as their carbon and energy source. However, it is clear that very many different carbon substrates can be converted into exopolysaccharides. These include amino acids, fatty acids, TCA cycle intermediates and hydrocarbons. The normal nitrogen sources employed are either ammonium salts or amino acids or, very occasionally, nitrate. The amino acids may serve as both carbon and nitrogen sources for some microbial species. Dinitrogen-fixing bacteria are almost all capable of exopolysaccharide production, and some at least may have potential industrial applications, although polysaccharide production and microbial growth may be best when a fixed nitrogen source such as NH_4^+ is added. Generally speaking, the structure of the polysaccharide produced by a microbial species is independent of the carbon substrate used. However, in some plant pathogenic *Pseudomonas* species, including *P. syringae* var. *glycinae*, the polysaccharide synthesised depends on the substrate used. If this is sucrose, the product is a levan; on glucose, alginates are formed. A switch in the type of polymer formed is apparently rare and is only possible when the microbial cells are capable of producing more than one type of exopolysaccharide. The bacteria which produce dextran are also unusual in that they will *only* synthesise that polymer when grown on sucrose as carbon substrate. The bacteria can grow on other substrates, such as glucose, but are unable to synthesise dextran. The use of different carbon substrates may affect the extent of acylation of polymers and may also affect the 'quality' of the polysaccharides as determined by some physical parameters.

Various ions are known to be required for the uptake of substrates or for

70

Table 6.1. *The effect of nutrient limitation on xanthan composition*

	Limiting nutrient				
	Carbon	Nitrogen	Magnesium	Potassium	Phosphorus
Uronic acid (%)	21	21	16	23	17
Acetate (%)	4.9	3.1	5.2	4.5	—
Pyruvate (%)	5.8	6.6	1.1	5.5	0.9

the synthesis of exopolysaccharides, and must therefore be present in adequate amounts in the medium. In most microbial species, K^+, Mg^{2+}, and to a lesser extent Ca^{2+}, are essential requirements for optimal growth. Other ions are also needed in smaller quantities and are usually present as components of other additions to the culture medium. Phosphate is the major anionic requirement of the microbial cells.

The balance between utilisable substrate and a limiting nutrient has frequently been shown to be of considerable importance for polysaccharide production. As the usual substrate is carbohydrate, this must be present in adequate amount or even in excess. During exopolysaccharide synthesis, it is neither necessary nor desirable to have a very high density of microbial cells. Media for polysaccharide production therefore tend to be low in nitrogen content relative to carbohydrate (e.g. 0.1–0.2% NH_4^+ salt; 2–3% glucose). Traditionally, nitrogen has been the limiting nutrient used in many studies on polysaccharide synthesis in either batch or continuous culture. However, for a number of bacterial species, sulphur (sulphate) or phosphate limitation can also be used. Potassium limitation can also be employed but often this fails to give good yields of exopolysaccharide because of the role of K^+ in substrate uptake. Under phosphate or magnesium limitation, in the case of xanthan (Table 6.1), there may be additional effects on polysaccharide acylation and on physical properties. When other polysaccharide-producing microbial species are used, it may be possible to study different limiting nutrients. Thus molybdenum limitation has been studied for its effect on bacterial alginate synthesis by *Azotobacter vinelandii*; its effect is presumably on dinitrogen fixation and it should perhaps be regarded in this bacterial species as being equivalent to nitrogen limitation.

The polysaccharide-producing microorganism may also either require or grow better in the presence of small amounts of amino acids and growth factors. These can either be added individually or in the form of materials such as casein hydrolysate or yeast extract respectively. The quality of a product such as xanthan and its final yield may also benefit from the addition of organic acids and TCA cycle intermediates, including α-ketoglutaric acid or citric acid. It has been suggested that this may lead to an improved metabolic balance between carbon flow from hexose substrate through the hexose monophosphate and Entner–Doudoroff pathways and oxidation through the tricarboxylic acid (Krebs) cycle.

Most polysaccharide-synthesising microorganisms are either aerobes or facultative anaerobes. In the latter, exopolysaccharide synthesis generally occurs only when the microorganism is grown aerobically. Thus aeration of the culture medium is an important requirement for polymer production in both prokaryotes and eukaryotes. However, not all microbial cells require maximal aeration for good production of the polysaccharides. A marked difference has been observed between *Rhizobium meliloti* and *Klebsiella aerogenes*, both growth and polysaccharide production in the former being optimal under lower aeration, whereas *K. aerogenes* requires vigorous aeration for maximal polysaccharide synthesis. In some microbial species, exopolysaccharide production is plentiful when the microorganisms are grown on solid media, but poor in the corresponding liquid media even under conditions of good aeration. It is not clear whether this reflects a need for micronutrients from the solidifying agent or reflects some facet of the aeration conditions required.

Uptake of carbon substrates

Clearly, microorganisms synthesising polysaccharides with high efficiency from carbon substrates, require effective systems for uptake of the substrate. This aspect has received relatively little attention despite its importance in the economics of polysaccharide production. A study in *Agrobacterium radiobacter* (Cornish et al., 1988a,b) has shown that three periplasmic proteins are derepressed during growth of the bacteria in glucose-limited continuous culture at low dilution rate. Two of these were high-affinity glucose-binding proteins, which were thought to be components of two independent, glucose-active transport systems. Similar transport systems have been reported in *Pseudomonas aeruginosa* and *Rhizobium leguminosarum*. In *A. radiobacter*, the levels of the two glucose-binding proteins increased following prolonged glucose-limited culture. The third protein was apparently not involved in glucose transport. However, all the substrate was converted into cell material and CO_2 under carbon limitation and no exopolysaccharide was produced. Growth under nitrogen limitation used the same periplasmic glucose-binding proteins. Transition from carbon to nitrogen limitation caused a rapid decrease in glucose uptake capacity, although the glucose-binding proteins were only slowly diluted out of the cells. A different strain, in which glucose uptake was less repressed than in the wild-type cells, had a relatively high concentration of the larger glucose-binding protein (relative molecular mass 3.65×10^4) and a higher rate of exopolysaccharide synthesis. Glucose uptake was a major kinetic control exerted over succinoglucan production in *A. radiobacter*.

In obligate and facultative anaerobes, including enterobacterial species, the commonest method of carbohydrate uptake is the phosphotransferase system. There have as yet been no similar attempts to relate carbohydrate uptake to polysaccharide synthesis in such bacteria.

Batch culture and growth phase

Exopolysaccharide produced in batch culture represents a mixture of molecules formed at different phases of the growth cycle. These may differ from one another in various respects; this is certainly true of xanthan grown in batch culture under different nutrient limitations. Differences can be seen both in acylation (Table 6.2) and in viscosity (Fig. 6.1). The reasons for the variations observed in both chemical composition and physical properties are not clear. There may be only a limited availability of some of the activated precursors at certain stages of growth. Thus acyl precursors in particular, such as acetyl CoA and phosphoenolpyruvate, may not be available in sufficient quantities at all phases of growth to permit complete acylation of the polysaccharide. This is less likely to be the case for sugar nucleotides. There is also the possibility that the demand on shared intermediates such as isoprenoid lipids cannot be satisfied and that the microbial cell allocates them according to well-defined priorities. Turnover of these molecules during polysaccharide synthesis *might* be more rapid, resulting in polysaccharide with lower molecular mass and reduced solution viscosity. As can be seen from Fig. 6.2, the viscosity of xanthan produced at 120 h in sulphur-limited batch culture was lower than that synthesised at 72 or 96 h. This may represent lower-molecular-mass polymer *or* the production of polysaccharide with missing components as cells age and lose viability. One must, however, regard these possibilities with caution. Different strains of the same microbial species may vary greatly when grown under the same physiological conditions.

Microorganisms differ with respect to the growth phase during which exopolysaccharide is produced. *Xanthomonas campestris* is unusual in that xanthan synthesis occurs throughout growth and into the stationary phase. Other microbial cells may behave very differently indeed. Bacterial alginate synthesis by *Pseudomonas aeruginosa* is mainly during the exponential phase of growth, whereas production of a heteropolysaccharide by another *Pseudomonas* species only started late in the exponential phase and continued to reach a maximum in the stationary phase.

6.3.   Industrial substrates

A clear distinction must be drawn between substrates used in laboratory experiments and those used for the industrial production of exopolysaccharides. In the laboratory, the main aim is to determine what substrates are utilised and what yields of biopolymers are obtained under different

Table 6.2. *Composition and viscosity[a] of xanthan from sulphur-, ammonium-, phosphorus-, magnesium-, glucose- and iron-deficient batch cultures of* X. campestris *S459*

Medium lacking	Time (h)	GlcA (%)	Acetate (%)	Pyruvate (%)	k^b (mPa s)	n^c
S	48	14.2	2.9	3.3	17	0.73
	72	15.4	4.1	3.8	122	0.47
	96	14.5	4.0	3.3	119	0.47
	120	18.4	4.1	2.7	78	0.52
NH$_4$	48	14.5	3.7	2.9	21	0.69
	72	15.9	4.5	3.0	134	0.45
	96	16.6	4.4	2.6	113	0.49
	120	15.3	4.4	2.2	76	0.54
P	48	12.9	2.7	—	9	0.79
	72	14.4	4.0	—	76	0.52
	96	16.7	4.0	3.6	82	0.52
	120	16.3	4.1	2.8	122	0.46
Mg	48	9.8	1.9	—	5	0.83
	72	11.9	4.0	—	94	0.51
	96	16.1	4.1	3.8	139	0.44
	120	15.6	4.0	3.2	96	0.49
Glc	48	—	—	—	—	—
	72	14.3	3.5	—	109	0.47
	96	16.4	4.2	3.8	128	0.45
	120	15.2	4.0	3.1	62	0.55
Fe	48	—	—	—	—	—
	72	14.9	4.2	—	120	0.46
	96	14.1	4.3	4.0	161	0.42
	120	14.8	3.9	3.0	101	0.49

Notes:
[a] 0.1% (w/v) Sodium-form xanthan in 9.6 mM NaCl (25 °C).
[b] k, Consistency index.
[c] n, Pseudoplasticity index (relative units).

physiological conditions. For such purposes, pure substrates such as glucose, fructose, sucrose or glycerol are frequently employed. In industrial production, the decision as to which substrate is to be used will be affected by its cost, but the yield and ultimate usage of the product may also affect the choice of substrate. For a high-quality product, the substrate must itself be of good quality and there must be no risk of carryover of impurities into the final product. As the normal substrates for polysaccharide production are carbohydrates, those most frequently employed for industrial purposes are starch, starch hydrolysates, corn syrup, glucose or sucrose (derived from cane or beet). When less pure products are acceptable, and the cost can effectively be reduced, cruder, cheaper substrates may be used. Some of

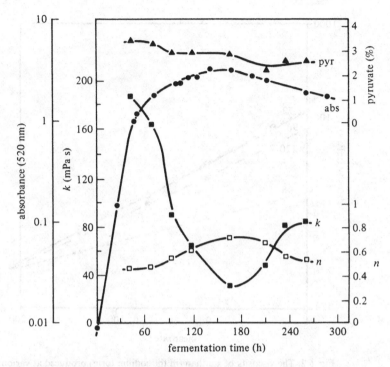

Fig. 6.1. Viscosity parameters and composition of xanthan grown in ammonium-deficient batch culture. Abbreviations: k, consistency index; n, pseudoplasticity index; pyr, pyruvate content; abs, culture absorbance (results of Tait, 1984).

these are waste products from other industries; for example, whey, cereal grain hydrolysates or dry milled corn starch may be suitable. As the production of exopolysaccharides generally requires a high ratio of utilisable carbohydrate to limiting nutrient such as nitrogen, the use of a crude substrate with high nitrogen content can greatly affect the product yield and may also alter the product quality.

Although most polysaccharide-producing microorganisms which have been commercialised so far produce optimal yields from carbohydrates as substrates, some species may be found to be more versatile. Indeed, *Xanthomonas campestris* can be grown on a very wide range of substrates, including both carbohydrates and protein hydrolysates. Bacteria utilising hydrocarbon substrates including methane and methanol have been proposed as sources of exopolysaccharides, but none of the products have yet been commercialised. For commercial production, the nitrogen source may be one of several proteinaceous products. These include yeast hydrolysate, soybean meal, distillers' solubles, cottonseed flour or casein

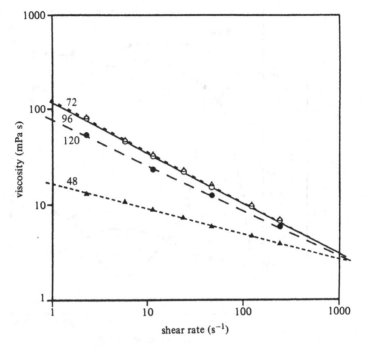

Fig. 6.2. The viscosity of xanthan (in the sodium form) produced at various stages of growth in sulphur deficient batch culture. Time after inoculation (hours) is shown for each growth stage.

hydrolysate. The choice may well depend on local availability as well as cost. The nitrogen source is used in concentrations of 0.05–0.2% by mass in the medium.

In addition to the carbon substrate and the nitrogen source, the medium used for polysaccharide production will contain sources of phosphorus, potassium and magnesium. Trace elements may also have to be added. The water used for medium preparation on an industrial scale will be from the public supply or other readily available source, or it may be condensate water if the polysaccharide fermentation plant forms part of an integrated production complex.

The production of dextran provides an interesting example of the specific requirements for industrial production of a microbial polysaccharide. The medium used contains inorganic salts with iron and manganese as important trace elements. The basic nutrients are 2% corn-steep liquor and sucrose. After sterilisation of the medium in a flow process, culture is performed at 25 °C with aeration and stirring. The pH drops, owing to the production of lactic acid, from *ca.* 7.0 to 4.0. During the process, the viscosity of the medium rises greatly, owing to dextran formation; after 3–4 days, a gelatinous mass of dextran is produced.

6.4. The fermentation process

The commercial production of exopolysaccharides is currently performed by batch fermentation, where the microorganisms are cultured under conditions that are both optimal for polymer production and economical in terms of substrate utilisation and conversion to product. Various continuous processes have also been described, but apparently none have as yet been adopted for industrial production of the microbial exopolysaccharides. Multistage processes have also been described. The first stage in one of these aims to yield cells under carbon limitation; thereafter, transfer to carbohydrate-rich media permits high yields of polymer. Alternatively, one process for xanthan production suggested the use of glycerol or xylose for the initial stage, substrates that permitted little polysaccharide synthesis but allowed relatively good growth of the bacteria. This was again followed by transfer to conditions in which sufficient carbohydrate substrate was available and polysaccharide yields were high.

In the design of any process for microbial exopolysaccharide production, it has to be remembered that the microorganisms differ in the phase of growth during which polysaccharides are produced. Synthesis of the polymers may be growth-associated, as in the case of *pullulan* from *A. pullulans* or bacterial alginate synthesis by *A. vinelandii*. Alternatively, it may occur only when growth has ceased. This is found for some *Pseudomonas* species and for the production of curdlan by *Alcaligenes faecalis* var. *myxogenes*. The production of xanthan occurs both during the growth of *X. campestris* and after cell growth is completed; the specific rate of xanthan synthesis is closely related with the growth rate in batch culture. It is maximal during exponential growth and minimal during the stationary phase. *Enterobacter aerogenes* strains tend to commence polymer synthesis late in the exponential phase of growth and continue production for some time in the stationary phase. Not all strains of the same species necessarily behave identically.

Continuous culture

Continuous culture appears to offer a number of advantages over batch culture for the production of polysaccharides. It provides high productivity from the fermenter, as there is not the necessity for frequent stopping and starting of cultures with attendant loss of production. The product is obtained at a uniform rate and the whole process can be readily geared to both the production and sterilisation of medium, and the recovery of the polysaccharide. The capacity of the fermentation vessel need not be as great as that used for batch production to yield the same amount of product. Unlike batch growth of the microorganism, there is no variability in the conditions employed and this should eliminate possible variations in product yield and quality.

In the laboratory, many studies on polysaccharide production have used continuous culture, generally with nitrogen or sulphur as the limiting nutrient. Single-stage continuous culture for xanthan production has also been suggested in the patent literature, but has apparently not been adopted as a commercial process. This can be understood when the *disadvantages* of continuous culture are considered. Strain stability may be very difficult to maintain under the physiological pressures applied in continuous culture. Variants with reduced or altered capacity for polysaccharide synthesis may readily emerge. If the variant can grow more rapidly under the selected culture conditions than the polymer-producing strain, it will quickly outgrow the original strain and cause loss of production. This is of course also true if contamination of the culture should occur. Again the contaminant is likely to outgrow the original species with loss of production. If the medium is relatively complex, the possibility of contamination is more likely; in simpler media, small numbers of contaminants may not pose such great problems. It is also difficult to maintain sepsis for the very long peiods, weeks or months, during which it may be wished to run continuous cultures. This will obviously be less difficult if the carbon substrate employed is one which is only utilised by a small number of microbial species, as would be true for methanol.

Under nitrogen-limited continuous culture for xanthan, approximately 25% of the carbohydrate substrate remained unused. The other major disadvantage of continuous culture is the relatively low product concentration in the fermentation broth. On the other hand, the lower viscosity of the fermentation liquor results in a lower energy requirement for stirring the fermenter. Carbon limitation (15 g l^{-1}) for xanthan production in continuous culture has also been suggested. The conversion efficiency of carbon source to polysaccharide was high and utilisation of the substrate was virtually complete. A further advantage of carbon-limited culture was the low BOD (biological oxygen demand) of the effluent produced by the process. The preferred dilution rate (D) was 0.034–0.05 h^{-1}. However, the viscosity and other characteristics of the product may be less satisfactory than those obtained by using nitrogen or sulphur limitation.

Strain selection

All commercial fermentation processes require the use of stable microbial strains. This is as true of polysaccharide production as of any other process. The strain must not degenerate during the batch process; as already pointed out, strain variation is even more likely if continuous culture is employed.

Considerable effort has gone into strain selection for the production of polysaccharides such as xanthan. As well as obtaining stable strains providing high polysaccharide yields, efforts have been made to remove unwanted microbial products. Thus there have been attempts to isolate mutants devoid of colour, lacking cellulase, etc. In the case of xanthan, the producer has one major advantage in comparison with other commercial fermentation products and other polysaccharides: *Xanthomonas campestris* only produces one exopolysaccharide and it does not use substrate for the synthesis of large amounts of intracellular storage products such as poly-β-hydroxybutyric acid (PHB) or glycogen. Nor is much substrate utilised for the production and excretion of extracellular enzymes. The conversion of substrate to extracellular polymer is normally very high indeed (over 70%). As far as is known, the industrialist wishing to manufacture xanthan is not faced with the problem of bacteriophage infection and it is not necessary to have available a series of producing strains differing in their phage-sensitivities. The production of other microbial exopolysaccharides is generally less efficient. Pseudomonads may convert much of the substrate provided into PHB; species from the Enterobacteriaceae are likely to form considerable amounts of intracellular glycogen. The physiological conditions employed for polysaccharide production are also those most likely to favour the formation of storage polymers. If the polysaccharide is from a *Rhizobium* species, the additional problem of multiple polysaccharide production is encountered. Clearly, the producer using any of these microorganisms must select a strain that is incapable of forming glycogen or any other unwanted products. Much less is known about fungal fermentations, in which the microorganisms may produce large amounts of the disaccharide trehalose (a storage product) as well as other unwanted products in addition to the desired polysaccharides.

In the case of microbial strains producing biosurfactants such as emulsan, the ability of mutants with enhanced extracellular polymer production to resist the antibacterial action of *cationic* surfactants has been utilised. Mutation for resistance to cetyltrimethylammonium bromide leads to increased production of the emulsan capsular material.

It is also possible to mutagenise strains and select mutants with improved polymer characteristics. These may vary from converting capsule-producing cells to slime-formers, thus making product recovery much simpler, to obtaining stable mutants which can synthesise polysaccharide of greatly increased viscosity. In either case the polysaccharide composition and structure remains unaltered.

Effect of growth rate

It is difficult to generalise about the effect of dilution rate on exopolysaccharide production. Each species, perhaps even each strain and limiting nutrient, has to be considered separately. Within a limited range

(D = 0.03–0.07h^{-1}), sulphur-limited cultures of X. *campestris* showed little effect on polysacchride acylation, although the rheological characteristics did alter (Fig. 6.2). The yield of exopolysaccharide was effectively constant within this range of dilution rates. The specific rate of polysaccharide synthesis in nitrogen-limited continuous culture has also been shown to vary relatively little for specific growth rates of 0.05–0.2 h^{-1} in X. *campestris* and in an exopolysaccharide-producing *Pseudomonas* species, respectively. In contrast to these results, yields of polysaccharide from *Agrobacterium radiobacter* in ammonia-limited continuous culture were highest at a dilution rate of 0.047 h^{-1} but the physical characteristics of the products were not reported.

Alginate production by A. *vinelandii* under phosphate limitation varied from a maximum at D = 0.15 h^{-1} to a level 24% lower at D = 0.05 h^{-1} and 5% lower at D = 0.25 h^{-1}. Using different limiting nutrients, at the same dilution rate, Jarman *et al.* (1978) observed very large differences in polysaccharide synthesis for this bacterial species (Table 6.3).

Strain maintenance

As well as being stable during the culture conditions employed to produce polysaccharides, the microbial strains must be maintained in such a way that they do not lose their desirable characteristics. Some can be regularly subcultured; others are held in a lyophilised state or frozen under nitrogen or at $-80\,°C$.

6.5.   Fermenter design

Although fermenter design is critical for the optimal production of exopolysaccharides, relatively few published reports relate to studies on the geometry and other parameters of the vessels used. Because of the high viscosity of the product in solution, one must ensure that mixing of the vessel contents is adequate. If it is not, both mass and heat transfer and the distribution of residence time (if the fluid in the vessel is not uniform, some portions will differ from others in their residence time) will be affected. Even in the best-designed fermentation vessel, shear stress is not uniform throughout. It has been suggested that three distinct regions exist. Near the impeller, there is high shear stress of the pseudoplastic liquid, resulting in micromixing. In the macromixing region, circulating flow is slow, as is shear stress. Finally there are regions of the fermenter which are effectively stagnant. The volumes of the respective regions depend on the tip speed of the impeller and on the design of the vessel. The low velocity in the parts of the fermenter furthest away from the impeller cause air bubbles to coalesce and lead to reduced oxygen transfer. Mixing of the highly viscous culture broths requires high power input and, ideally, impeller blades closely

Table 6.3. *The effect of growth-limiting nutrient on exopolysaccharide production by* Azotobacter vinelandii. *Bacteria were grown in continuous culture at* $D = 0.15 \, h^{-1}$

Limiting nutrient	Exopolysaccharide produced (mg C g cell^{-1} h^{-1})
Molybdate	10.5
Phosphate	8.9
Iron (Fe^{2+})	8.1
Sulphate	7.8
Carbon (sucrose)	7.7
Nitrogen (N$_2$)	6.8
Potassium	5.0

Results from Jarman (1979).

similar in dimensions to the radius of the culture vessel. If there is too large a space between the impeller and the wall of the vessel, much of the fluid may remain unmobilised; as the viscosity increases, the volume of liquid set in motion by the impeller decreases. Impellers with large paddles fail to produce a satisfactory pattern of mixing. Much better results are obtained through the use of a double helical or a screw mixer. In practice, two mixing systems may be required: a slow ribbon mixer and a second turbine agitator to provide intense local turbulence and resultant good micromixing. The presence of numerous baffles, cooling coils and measurement probes all impede the flow of the viscous fermentation fluids and result in poor mixing. The ideal shape, from the point of view of efficient mixing, would be a spherical vessel, but in practice a low cylindrical form will probably be adopted. The types of fermenter configuration chosen for polysaccharide production clearly require careful choice but are apparently still closely guarded industrial secrets.

In the selection of reactors for microbial polysaccharide production, most industrialists probably use conventional aerated tanks with mechanical agitation, despite the problems of oxygen transfer and of heat removal. Air-lift fermenters can be ruled out because of the difficulty in achieving adequate mixing by this method. The design requirements of the vessel demand a balance between the impeller velocity head and the impeller flow. The former regulates the formation of gas bubbles; the latter controls the homogeneity of the liquid throughout the fermenter and the distribution of the air bubbles, as well as the rate of heat transfer. Impeller design should provide good dispersion of the air bubbles by obtaining sufficient turbulence in the fermenter contents. Stirring with straight blades yields high turbulence and high shear rate close to the impeller, but relatively low values in the remainder of the liquid.

As microbial exopolysaccharides are typically highly viscous and non-Newtonian in aqueous solution, very great changes occur in the rheology of the contents of the fermentation vessel during microbial growth and polymer synthesis. The microbial cells add only slightly to the viscosity of the medium if bacteria are being used, whereas fungal mycelia will increase viscosity. From a starting viscosity which is negligible, the culture fluid will change to high viscosity as polysaccharide is excreted into the medium. Associated with this will be changes in heat and oxygen transfer. These changes are liable to be even larger if fungal mycelia are present in addition to exopolysaccharide. However, if the polysaccharide is poorly soluble and precipitates from solution, viscosity increases will be much smaller.

Laboratory-scale fermentation vessels are usually cooled adequately by small water-circulating systems. On any larger scale, cooling is necessary to remove the heat generated through microbial metabolism and through the stirring process. As internal cooling coils are undesirable, any cooling system must operate through the walls of the vessel with coils placed outside. Such a system is probably satisfactory for vessels of capacity up to 50 m^3, provided that the temperature of the coolant is low enough.

6.6.   Product recovery

The aims of the polysaccharide recovery process include concentration of the fermentation broth to a form which is stable, easy to handle, and can be readily redissolved for the chosen application of the polymer. During recovery, purification of the product and removal of the associated non-polysaccharide components of the fermentation liquor should be obtained. Further, undesirable activities, which may be enzymes or components leading to colour, flavours or odours, should be destroyed or removed.

Treatment before recovery

Polysaccharide-containing fermentation broths may be processed before polymer recovery, either to enhance the quality of the product or to remove some unwanted characteristic. Pasteurisation of fermentation broths serves a number of purposes. The microbial cells are killed. They may also be partly lysed as a result of pasteurisation and thus reduce the amount of particulate material present in the broth. Alternatively, after the heat treatment involved in pasteurisation, the cells may be more sedimentable and more easily removed from the product. Pasteurisation also serves to inactivate enzymes present in the broth. Thus xanthan is freed of cellulase and protease as well as other thermolabile enzyme activities. This is particularly important when xanthan may be admixed with other polymers such as carboxymethylcellulose, which would be degraded by the cellulase.

The temperature must be carefully controlled to ensure that no degradation of the exopolysaccharide product occurs. Thus, the pH should be such that it will not enhance polymer hydrolysis or the loss of acyl groups which may contribute to the desirable rheological properties of the material. In the case of xanthan, heat treatment at pH 6.3–6.9 actually *enhances* the viscosity of the polysaccharide solutions at low shear rates. There is also some disruption of the cells of *X. campestris* present in the broth.

Removal of the microbial cells from the fermentation broths may be possible through the use of centrifugation or filtration. However, even with the addition of filter aids, filtration is rendered difficult by the high viscosity and non-Newtonian characteristics of the broth products. This also makes cell removal by centrifugation extremely difficult as well as being very energy-intensive.

As broth solutions containing polysaccharides are so viscous, they are normally diluted prior to filtration. It may also be possible to lower the viscosity by heating the polymer solution above the transition temperature before filtration. Clearly, some polysaccharides show little change in viscosity with increased temperature, so it is not practicable to do this with all preparations. Heat treatment can most easily be achieved in an integrated process in which the fermentation broth at the incubation temperature is heated during pasteurisation, then filtered while still at a relatively high temperature.

The industrialist wishing to recover exopolysaccharide from fermentation broth is faced with the problem of recovering product from an aqueous medium with low solid content. For some purposes, removal of cell debris may be unnecessary, but for others he or she may need to introduce some form of treatment to remove the microbial cell material. The presence of non-viable bacterial cells (after pasteurisation of the fermentation broths) can present adverse effects if the polysaccharides are to be used for either oil recovery or food use. Owing to the high viscosity of the fluids, filtration may be extremely difficult. It may, however, be possible to treat the pasteurised fermentation liquor with enzymes to degrade the walls of the microorganisms and thus effectively cause lysis and destruction of the cells. It has been proposed that proteases should be employed for this purpose in the preparation of xanthan solutions for use in enhanced oil recovery. Such a treatment certainly reduces the amount of particulate material present and increases the injectivity of the polysaccharide solutions. Protease treatment is probably more effective if performed after pasteurisation, than when viable cells are present. Another suggested procedure is to treat the material with an acid or neutral protease for a period of up to 10 h at temperatures up to 60°C, then raise the pH to a value of 8–13 and hold the preparation at ambient temperature. This alkaline treatment improves still further the filtrability of the polysaccharide solution. Finally the pH is lowered to neutrality. Such a procedure removes the non-viable bacterial cells and also clarifies the polymer solution considerably. It does, however, lead to loss of ester-linked substituents such as O-acetyl groups. If the polysaccharide is to be precipitated with organic solvents, the addition of an electrolyte is frequently practised, as this greatly increases the amount of polymer recovered.

Chemical treatments provide an alternative method of cell lysis. The chemical agents must, however, be handled carefully or the polysaccharide product may be degraded. Cells can be lysed by treatment with alkali, but this will also deacetylate the polymer and remove any other ester groups present. It is thus necessary and desirable for a polysaccharide such as gellan, from which the esters must be removed to yield a gel-forming product. For production of the deacetylated form of the polysaccharide, the fermentation broth is heated for 10–45 min at 90–100 °C at a pH, adjusted with sodium hydroxide or sodium carbonate, of about 10. However such treatment will result in the loss of acetyl groups from xanthan; this may be undesirable, although it improves the synergistic gelling of the product with galactomannans. Hypochlorite treatment can also be used for cell lysis. It has the added advantage of bleaching the product and inactivating extracellular enzymes, but is liable to cause some oxidative depolymerisation of the polysaccharide if the process is not very carefully controlled. Xanthan broth may also be treated with dialdehyde (polyglyoxal) to improve the dispersability of the product on re-solution after drying.

The recovery of scleroglucan differs slightly from that of bacterial products. The culture fluids are first heated to inactivate glucan-degrading enzymes and to kill the microbial cells. The culture is then homogenised to detach the polysaccharide from the fungal mycelium. The resultant viscous fluid is first diluted, then filtered to remove the particulate material. After the filtration process, the clear solution is concentrated prior to precipitation with isopropanol.

Recovery

Two major methods are used for the recovery of the exopolysaccharides from fermentation liquors. The polymer can be precipitated by the addition of polar organic solvents, which are totally miscible with water. The chemicals most frequently used are the lower alcohols such as methanol, ethanol or propan-2-ol (or acetone). As well as lowering the polymer solubility and permitting phase separation, they assist in the removal of colour and low-molecular-mass fermentation products and medium components. Propan-2-ol is often the alcohol used, but the choice may depend on various factors (for example, chemical firms may have large amounts of readily available methanol derived from natural gas or other sources). Propan-2-ol also complies with the legislative requirements for food-grade polysaccharides. Precipitation with lower alcohols is applicable to most microbial exopolysaccharides, whether they are polyanionic or neutral polymers. The effects are complex; neutral polymers are rendered less soluble in a single phase medium of changed composition through lowered affinity of the hydrophilic portions of the polysaccharides. Coacervation may even occur as an intermediate stage between solution and precipitation. In this intermediate stage, the water is effectively immobilised within the polymer aggregates. The polyanionic materials may exhibit additional effects, including enhanced binding of cations. The presence of the cations also reduces the amount of organic solvent required; salts such as KCl are added to xanthan broths for this purpose. After precipitation of the polysaccharides, the solvents are recovered by distillation.

The precipitated product must be freed from excess liquid by centrifugation or pressing. It is also possible to apply some chemical treatments at this stage if necessary. Final drying is achieved as a continuous process under vacuum or with inert gas to reduce the dangers associated with flammable organic solvents. As with pasteurisation, the conditions have to be carefully chosen to ensure that no product degradation or discoloration occurs. Too rapid drying may reduce the ease of solubility of the final product or cause excessive microgel formation on re-solution. The dry product is finally milled to a designated mesh size. This is chosen to ensure ease of dispersion and solution in aqueous solvents, as well as facilitating the handling of the solid product. The material leaving the production facility is then a fine buff or off-white powder. Purer preparations are white. Typically, such materials contain about 10% moisture and 5–6% ash. The polysaccharide must be stored in a cool, dry atmosphere to prevent adsorption of moisture and resultant biological degradation. The dry powder then remains stable if stored at temperatures below 25 °C.

Alternatively, ultrafiltration or reverse osmosis can be applied to reduce the water content of the product. Either method has its particular advantages and disadvantages, but each adds considerably to the cost of the product. Although there is renewed interest in developing improved methods of product recovery, the techniques available to the industrial polysaccharide producer are still very limited. It has also to be remembered that the product is a bulk chemical of only intermediate value (possibly US$10 000 t^{-1} for xanthan) and may have to compete in the market place against many other natural and synthetic polymers. The production of polysaccharides introduces the additional problem of the very high broth viscosity, not found with other microbial products.

The alternative recovery method to solvent precipitation, is filtration. This has the advantage that, at the same time as the solution of polysaccharide is being concentrated, there is removal of low-molecular-mass fermentation products and unused medium components. There is relatively low energy requirement ($1–12$ kWh m^{-3} of water removed) and the solution is only held for a brief time during the process, thus minimising any possible biological degradation. There is no requirement for solvent recovery and the water can if necessary be recovered for further fermentation cycles by reverse osmosis. This has the great advantage that there is little material of high BOD to be disposed of. Although the polysaccharides are not thermolabile, the process is carried out at ambient and not elevated temperatures. The design of modern cross-flow filtration techniques utilises the shear-thinning effects seen during the pumping of microbial polysaccharides, most of which possess pseudoplastic rheological characteristics. Consequently, the high shear achieved within such filtration systems permits the concentration of products in solution with viscosities up to 4000 cP. It is also possible to combine enzyme clarification with the filtration procedures.

A small number of exopolysaccharides may be recovered from solution by direct addition of acid or alkali to the cell-free culture supernatant. In the case of curdlan, addition of alkali results in the precipitation of the polysaccharide as a gel which can readily be recovered. Similarly, *some* bacterial alginates (those from *Azotobacter vinelandii*), which resemble the marine algal product in possessing contiguous sequences of L-guluronosyl residues, can be rendered insoluble through the addition of acid to the culture supernatant.

If the polysaccharide is in the form of a concentrate produced by filtration, it is likely to be obtained as a thick colourless gel containing about 4–8% solids. This must contain suitable antimicrobial agents (formaldehyde or azide) to inhibit microbial growth and biological degradation. Even so, some deterioration may occur if the product is kept for a prolonged period prior to use. The mechanism of this is not understood but it appears to affect material stored at ambient temperature or even in the cold.

Although several other recovery methods are possible, they find relatively few applications in the production of microbial polysaccharides. It may be possible to precipitate the polysaccharides with multivalent cations. Such a process has indeed been suggested for xanthan production, involving precipitation of the insoluble calcium or aluminium salts. The pH must be carefully controlled during addition of the salt, either at pH 10 for the calcium salt or between pH 3.5 and 4.5 for the aluminium salt. There may subsequently be problems in the removal of the salts because of the very high affinity of some polysaccharide structures for certain ions. Even the addition of sequestering agents may not provide a method for complete removal of the salt used. The use of methods such as these does not appear to have been widely adopted; however, in the specific example of bacterial alginate produced by *Azotobacter vinelandii*, the close structural similarity of this polymer to the traditional algal polysaccharide suggests that similar recovery methods can probably be applied if the process is commercialised. Because of the very high affinity of alginates for Ca^{2+} and Sr^{2+}, the addition of these ions provides a ready means for alginate recovery.

An alternative to precipitation with solvents or multivalent ions has utilised the fact that many microbial polysaccharides are polyanionic in nature. They can therefore form insoluble complexes with quaternary ammonium compounds. This property is quite widely utilised on a laboratory scale for polysaccharide recovery and also forms the basis for an ingenious method of polysaccharide quantification. It has also been proposed as a large-scale technique for polysaccharide recovery after removal of the bacterial cells from culture supernatants. The broth is adjusted to a suitable viscosity and 2% KCl added, followed by commercial cetyltrimethylammonium chloride. In this recovery process the concentration of the electrolyte is critical; insufficient salt results in a gel-like precipitate, and excess inhibits precipitation. The precipitated complex of quaternary ammonium compound and polysaccharide can then be treated with methanol to remove the precipitant, filtered and vacuum-dried. The disadvantage of such a method is the requirement for recovery of both the methanol and the relatively expensive quaternary ammonium compound. Thus, although pilot plant processes of this type have been described, the method has not found general acceptance for commercial recovery of microbial polysaccharides. It is possible that it might have some specialised application in the small-scale production of a particularly valuable polysaccharide.

Spray-drying is widely used in the food industry but has not found significant applications in polysaccharide recovery. It can be applied on the laboratory scale but suffers from the disadvantage that the initial culture broth is often very viscous and normally still contains large amounts of salts and has relatively low polysaccharide concentration. Spray-drying also involves a relatively large energy input and plant cost when applied on an industrial scale. Some preliminary purification is desirable before spray-drying; otherwise, impurities including salts, microbial cells and metabolic products will be present. Also in the presence of such impurities, the product is liable to be discoloured by the heating involved in the process. In the laboratory, lyophilisation is frequently employed, but this is neither practical nor cost-effective on a very large scale.

Product treatment

Most of the polysaccharides are used in their macromolecular form. An exception is dextran. If the products of fermentation are of high molecular mass, they are hydrolysed at 100–105 °C with hydrochloric acid. The extent of hydrolysis may be followed by measurement of viscosity, the reaction being stopped by addition of sodium hydroxide. The product is clarified and fractionated by methanol precipitation.

Further reading

Atkinson, B. & Mavituna, F. (eds) (1983). Downstream process engineering and product recovery. In *Biochemical engineering and biotechnology handbook*, pp. 890–931. Macmillan, London.

Saier, M. (1985). *Mechanisms and regulation of carbohydrate transport in bacteria*. Academic Press, Orlando.

Smith, I. H. & Pace, G. W. (1982). Recovery of microbial polysaccharides. *Journal of Chemical Technology and Biotechnology* **32**, 119–29.

Vincent, A. (1985). Fermentation techniques in xanthan gum fermentation. *Topics in Enzyme Function and Biotechnology* **10**, 109–45.

7   Genetics, control and regulation of exopolysaccharide synthesis

7.1.   Introduction

Although some of the earliest studies on bacterial transformation utilised as a model system polysaccharide production in *Streptococcus pneumoniae* and its relation to virulence, further progress in studying the genetics of polysaccharide synthesis has taken some considerable time. Much effort was applied to studies on the genetics of *colanic acid* synthesis in *E. coli* and *Salmonella typhimurium*, but this has proved to be a very complex system with numerous regulatory mechanisms. The complexity may perhaps be, at least in part, related to the relatively large hexasaccharide repeat unit of this exopolysaccharide and the ability of most bacteria synthesising it to produce more than one extracellular polysaccharide. Recently, however, detailed knowledge of the genetics of exopolysaccharide synthesis has derived from various systems. The interest in xanthan as an industrial product from *X. campestris*, and in the bacteria *per se* as plant pathogens, has prompted their study. *Rhizobium* species, as well as producing at least two different polysaccharides of potential industrial interest, have received attention because of their symbiotic relationship with leguminous plants and the associated bacterial fixation of dinitrogen. Alginate production by *Pseudomonas aeruginosa* has been studied because of the correlation between polysaccharide secretion and the infection of cystic fibrosis patients. Finally, a range of (mainly pathogenic) Gram-negative bacterial species have been examined, *E. coli* strains being used for a number of studies. All this information enables us to see some common aspects in the genetic control and regulation of exopolysaccharide synthesis; the concept of a '**cassette**' of biosynthesis genes unique for each polysaccharide, first conceived in exopolysaccharide-synthesising *E. coli* by Boulnois and his colleagues, may well be at least partly applicable to many, if not all, exopolysaccharide-synthesising bacteria. Unfortunately, virtually nothing is yet known about the genetics of exopolysaccharide production in eukaryotic microorganisms.

7.2.   Common genetic features in polysaccharide synthesis

One of the most intriguing aspects of genetic control of exopolysaccharide synthesis derives from study of a range of *E. coli* strains which form various K antigens including the K1 serotype composed of sialic acid. In each of these *E. coli* strains, similar organisation of the gene cluster controlling K antigen (exopolysaccharide) synthesis was observed. A region of approximately 17 kilobases (kb) of DNA formed three functional segments. The first region, extending to about 9 kb, appeared to code for genes functioning in the translocation of the completed polysaccharide to the bacterial surface. Transposon insertions in this region yielded cells in which the polysaccharide appeared in the periplasmic space. A second region of 5 kb consists of the genes responsible for the enzymes involved in biosynthesis of the sugar nucleotides specific to the polymer, as well as specific transferases and polymerase. Mutants in this region were exo⁻, failing to produce any exopolysaccharide. The functions of the third region were less clearly defined, being apparently involved in modification of the exopolysaccharide after it had reached the cell surface. This might include attachment of the polymer to the cell surface through a terminal linkage including the sugar ketodeoxyoctonic acid (Echarti *et al.*, 1983). Transposon mutagenesis in this region caused intracellular accumulation of the polysaccharide in a form differing from the mature polymer. Mutations in regions 1 and 3 prevent expression of exopolysaccharide at the bacterial surface but do not inhibit the enzymic reactions of polymer synthesis. On the other hand, one phenotype, observed from mutants in region 2, is the absence of polysaccharide production *in vivo*. Analysis of the genes of region 1 in different *E. coli* strains indicated DNA sequence homology; analysis of the proteins encoded by the homologous DNA sequence revealed sets of similar polypeptides. Complementation of the functions in regions 1 and 3 was possible, indicating that modification of exopolysaccharide after polymerisation, and transport of the mature polysaccharide from the site of synthesis, are mechanisms common to the different *E. coli* strains and are independent of polysaccharide structure (Fig. 7.1) (Roberts *et al.*, 1988). On the other hand, the central cassette of genes responsible for biosynthesis is unique to each bacterium and clearly will vary in sequence and size depending on the size of the repeat unit and the quantity of genetic information required. In other polysaccharide-synthesising bacteria, the genetic organisation may be different; this type of genetic control may be limited to certain types of capsular material formed by *E. coli* strains or to species of Enterobacteriaceae. Recently, some homology has been demonstrated between genes involved in polysaccharide synthesis in *E. coli* and *Klebsiella aerogenes*.

Common features have also been shown to be involved in the translocation of *E. coli* polysaccharides to the bacterial surface. A 60 kDa periplasmic protein, the product of the *kpsD* gene, probably represented a common feature of several different capsular serotypes. This protein resembles other periplasmic proteins in being synthesised as a precursor with a leader sequence which is removed prior to appearance of the protein in the periplasm. The protein is one of five with molecular masses 80, 77, 60,

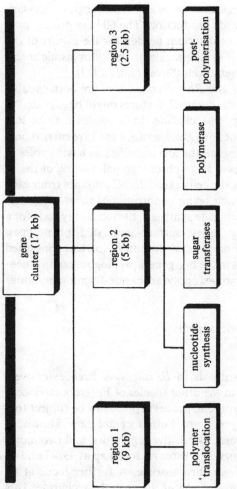

Fig. 7.1. The cassette of *E. coli* genes thought to control polysaccharide synthesis.

40 and 37 kDa respectively, which are coded by an 11. 6 kb region necessary for the expression of polysaccharide production. The 60 kDa protein may thus form part of a multicomponent system needed for the export of the surface polysaccharides from the synthetic site at the cytoplasmic membrane, to the exterior of the bacterial cell (Silver *et al.*, 1987).

As *Agrobacterium tumefaciens* and *Rhizobium meliloti* are both capable of synthesising succinoglycan, some common features might be expected in the genetic control of this exopolysaccharide. In *R. meliloti*, three loci involved in polysaccharide production (*exoA*, *exoB*, *exoF*) are carried on a plasmid, whereas in *A. tumefaciens* **all** the loci identified as having roles in exopolysaccharide synthesis appear to be chromosomal. Curing of the Ti plasmid in *A. tumefaciens* leaves the cells exo$^+$. *ExoC* mutants from each bacterial species were similar, each being pleiotropic, slow growing and partially defective in exopolysaccharide synthesis; the locus may code for a cell surface structure affecting many surface-associated phenotypes. Despite the difference in the genotypic distribution of the loci concerned with exopolysaccharide synthesis in the two genera, analogous complementation groups are found in any non-exopolysaccharide-forming mutants.

7.3.   Specific systems of genetic control

Escherichia coli

The genetics of colanic acid synthesis in *E. coli* have been extensively studied. In *E. coli*, and possibly in the other species of Enterobacteriaceae that produce colanic acid, the synthetic process appears to be subject to a particularly complex regulatory system. Four loci are now known to function in this: two, *rcsA* and *rcsB*, as positive regulators; and two more, *lon* and *resC*, as negative regulators. The three *rcs* loci map at 43, 47 and 48 minutes respectively on the *E. coli* chromosome. A further locus at 63 minutes is concerned with overproduction of the exopolysaccharide. This locus is linked to *serA* and thus resembles the *kps* genes involved in the synthesis of several *E. coli* capsular (K) antigens (Table 7.1).

The rcsA gene product, an unstable protein of 27 KDa, has a half-life of 5 min in *lon*$^+$ bacteria and 20 min in *lon*$^-$ mutants. Availability of this product is a limiting factor for capsule synthesis; increased gene dosage increases expression of the *cps* genes. In contrast, the product of the *rcsB* gene is stable in *lon*$^+$ bacteria. Multiple expression of the *rcs B* gene in rcsC mutants is lethal, possibly because of accumulation of intermediates in colanic acid biosynthesis. The *rcsC* locus is more complex. *E. coli* K12 *lon* mutants overproduce colanic acid but are defective in an ATP-dependent protease and are UV-sensitive. The sensitivity of the mutant cells to DNA damage results from a defect in proteolysis, which permits an unstable cell-division product (sulA) to accumulate. It was therefore predicted that the

Table 7.1. *Regulation of colanic acid synthesis in*
E. coli

Regulator		Map site (minutes)	Product M_r
rcsA	+ve	43	27kDA
rcsB	+ve	47	26kDa
rcsC	−ve	48	100kDa
lon	−ve	63	
rcsA[a]	+ve	ND[c]	25kDa
rcsA[b]	+ve	ND[c]	23 kDa

Notes:
[a] From *Erwinia stewartii*.
[b] From *Klebsiella aerogenes*.
[c] ND, map site not determined.

target of the *lon* protease, an unstable positive regulator of colanic acid biosynthesis, would normally be limiting in the bacteria. The target polypeptide was later revealed to be an unstable protein of 27 kDa, the product of *rcsA* (Torres-Cabassa & Gottesman, 1987). A limiting factor for colanic acid synthesis is the availability of the *rcsA* gene product. Doubling gene dosage significantly increases expression of *cps* genes. Mutation in *lon* increases synthesis of colanic acid because of the increase in stability of the *rcsA* gene product. The *lon* gene is responsible for negative regulation of at least six structural genes (*cps*A–F) at the transcriptional level. These genes are also subject to regulation by *rcsA, B* and *C*.

Two genes cloned from the chromosome of *K. aerogenes* activated expression of synthesis of colanic acid in *E. coli* (Allen *et al.*, 1987). It was suggested that one of the gene products acted as a positive regulator of colanic acid formation and that its activity was subject to regulation by the *lon* protease. The second locus could bind preferentially to a negative regulator of exopolysaccharide biosynthesis; no polypeptide coded by this locus was identified. Homologous gene sequences were found in the genomes of other exopolysaccharide-producing *K. aerogenes* but not in *E. coli*.

It is thus possible that a considerable number of exopolysaccharide-forming bacteria in the Enterobacteriaceae utilise proteins resembling the *rcsA* gene product to regulate exopolysaccharide synthesis.

In *Erwinia stewartii*, in which there is a direct relationship between exopolysaccharide production and plant pathogenicity, a positive regulator shows considerable similarity to *rcsA* in *E. coli*. In *E. stewartii*, the gene promotes transcription of two operons (*cps*) involved in exopolysaccharide synthesis. Like *resA* in *E. coli*, it codes for a protein of 25–27 kDa; it has a short half-life in *E. coli*, but is more stable in *E. coli lon* mutants. It also complements *E. coli rcsA* mutants; the *E stewartii rcsA* gene product increased *cps* expression in *E. coli*. There is thus a high degree of homology with regard to exopolysaccharide production in these two bacterial species. A large region of the Erwinia chromosome linked to *his* and involved in polysaccharide synthesis is analogous to the six *cps* genes also located near *his* in *E. coli* and regulated by *rcsA* (Dolph *et al.*, 1988). Differences do exist between the exopolysaccharide-synthesising systems in the two bacteria as fucose is present in colanic acid only, although both polymers contain glucose, galactose and glucuronic acid.

Another *E. coli* system which has received study is that responsible for the synthesis of sialic acid in K1 strains. Genetic studies demonstrated the involvement of a 9 kb region comprising at least two or three groups of genes. One group is responsible for the biosynthesis of the polysacchride capsule and the others for translocation and transport. Some DNA homology with other *E. coli* strains was noted in the genes controlling those aspects of the process that were not specific to polysaccharide structure; i.e. transport and related processes were independent of the polysaccharide structure. A group of polypeptides of 38, 60, 42, 49, 74 and 30 kDa were common gene products. Elimination of certain of these polypeptides resulted in cells containing periplasmic polysaccharide. Loss of the 38 kDa protein caused the bacteria to form blebs, which stained with ruthenium red but did not react with antibodies. This protein may therefore be concerned with the final stages of polymer export to the external environment. The strains showing homology included strain K92, which also secretes a sialic acid polysaccharide, and strains K7 and K100, which produce polymers of different composition. Little or no homology was found with *Neisseria meningitidis* group B, despite the chemical and immunological identity of the sialic acid from these bacteria. However, transfer of a segment of genetic material from *Neisseria meningitidis* containing a promoter permitted expression of sialic acid synthesis in *E. coli*. In addition to this, expression of sialic acid synthesis in *Salmonella typhimurium* was obtained following introduction of a plasmid coding for the *E. coli* K1 polysaccharide. There is thus probably considerable homology of DNA in **closely related bacterial strains** but less in other species that are remote in systematic terms.

Pseudomonas aeruginosa

Alginate synthesis differs from the manufacture of other bacterial exopolysaccharides as indicated in Chapter 5. It is therefore of interest to determine whether differences also exist at the genetic level. Problems have also been encountered because of the apparent low levels of a number of the enzymes involved in precursor synthesis, the presence of other activities interfering in the enzyme assays and the failure to obtain a satisfactory cell-free system. Another problem has been the difficulty in obtaining stable

Table 7.2. Pseudomonas aeruginosa *genes involved in alginate synthesis*

Gene	Enzyme	Product
algA	phosphomannose isomerase	56 kDa
algB	phosphomannomutase	(58 kDa)[a]
algC	GDP–mannose pyrophosphorylase	
algD	GDP–mannose dehydrogenase	48 kDa

Note:
[a] This may be the product of an *alg* gene regulated by *algB* rather than the direct product of *algB*.

alginate⁻ mutants, owing to the instability of the mucoid phenotype. Progress has thus been slower than that for other polysaccharide-producing systems. Despite this, considerable effort has gone into the study of the mechanisms of genetic regulation of alginate synthesis.

Four of the genes (A–D) in alginate synthesis control the enzymes yielding the activated precursor GDP–mannuronic acid (Table 7.2). Most mutants affecting alginate production are located at 45 minutes on the *P. aeruginosa* chromosome, near the *argF* gene. An exception is the *algB* gene in the 21 minute region; *algR* is close to *argH* at 19 minutes. Some but not all of these genes involved in alginate synthesis are found near loci involved in carbohydrate metabolism. That may be significant, given the differences in derivation of the hexose moiety in alginate compared with other bacterial polysaccharides. It is clear that there is transcriptional control of alginate production in mucoid *P. aeruginosa* in the *algA* locus, whereas the *algD* locus is not subject to such control. The *algD* promoter is under positive control by the *algR* gene. The *algR* gene resembles several others which have been characterised in *E. coli* in being environmentally responsive, and indicates that it is part of a complex regulatory system which exerts environmental control over alginate synthesis.

The *algD* promoter segment is rich in AT bases, a feature also seen in strong *E. coli* promoters. It also contains direct repeats, upstream of the transcription start. The *algD* promoter also displays osomolarity-dependent action, an aspect that may be significant in relation to the origin of alginate-synthesising *Pseudomonas* from cystic fibrosis patients.

The genes in one region of the *P. aeruginosa* chromosome appear to be mainly concerned with early steps in alginate synthesis (prior to polymer formation). Another region may be involved in polymerisation, secretion and transport, acetylation and epimerisation (Wang *et al.*, 1987). This second region carries a tight clustering of several genetic loci involved in alginate biosynthesis. It contains the *algB*, *algC* and *algD* genes, together with others of unknown function. Transcription of the known *alg* loci is unidirectional, proceeding from *algD* to *algB*. Another two *alg*-related polypeptides are coded by the *alg* gene cluster and the genes appear to be transcribed in the same orientation as part of a single regulatory unit. As with other polysaccharide-synthesising systems, the genes identified so far for alginate synthesis are not located in the same order as the biochemical steps for which they code. Although an alginate epimerase has not yet been isolated from *P. aeruginosa*, a gene thought to code for such activity has been cloned.

Two further tightly linked genes, *algS* and *algT* at about 68 minutes, appear to be involved in the interconversion of alginate$^+$ and alginate$^-$ strains. Evidence has been presented to suggest that the *algS* gene functions in *cis*, activating the *trans*-active *algT* gene, and that the product of the latter is necessary for alginate production.

Xanthomonas campestris

Among studies on genetic aspects of polysaccharide synthesis, recent reports on the xanthan system have revealed much of interest as well as providing mechanisms for the production of increased rates and yields of polysaccharide. It has also proved possible to obtain xanthan altered in its acetylation, pyruvylation and side-chain oligosaccharides. In *X. campestris*, mutations have been found in a 13 kb DNA sequence, which appears to control polysaccharide biosynthesis. Among the genes linked to this cluster is one involved in pyruvylation of xanthan. Plasmids containing this gene could be used to enhance the pyruvylation of xanthan by 45%; other genes could increase polysaccharide synthesis by about 10%. It has been suggested that not all the genes essential for xanthan synthesis are clustered; one locus has been subdivided into three distinct complementing regions and a second locus contained two complementation groups. Libraries of DNA fragments from wild-type *X. campestris* have been cloned into *E. coli*, through the use of a cosmid vector with a broad host range, and subsequently transferred by conjugal mating into non-mucoid *X. campestris* mutants. Insertion of multiple copies of cloned genes into the wild-type bacteria generally had no noticeable effect and in some cases even decreased the polymer yield. In a few of the wild-type recipients, increased xanthan accumulated; in one, the rate of xanthan accumulation under defined conditions was twice that of the control. The increase in total polysaccharide synthesised never exceeded 20% and the highest conversion of carbon substrate to polysaccharide was 80%. These experiments were performed on a small scale in the laboratory and it is not clear whether (i) the recombinant strains are stable following prolonged use in the laboratory; and (ii) the increased rate or yield of xanthan can also be observed during industrial-scale production.

The most comprehensive study on *X. campestris* has demonstrated a cluster of 12 genes involved in xanthan production in a 16 kilobase region of

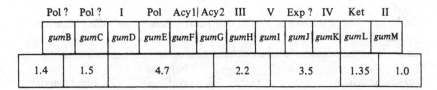

Pol?	Pol?	I	Pol	Acy1	Acy2	III		V	Exp?	IV	Ket	II
gumB	gumC	gumD	gumE	gumF	gumG	gumH	gumI	gumJ	gumK	gumL	gumM

1.4	1.5	4.7		2.2		3.5		1.35	1.0

Fig. 7.2. The restriction map of *Xanthomonas campestris*. The *Bam*HI restriction map (bottom) indicates the order and the approximate size (in kilobases) of the fragments of the 16 kb of DNA of the xanthan gene cluster. The genetic map (centre) indicates the twelve separate xanthan genes, designated *gum* B to *gum* M. The biochemical functions (top) indicate the enzymic activity identified for each gene: I–V, transferases I–V; Acy, acetylases (1 and 2); Ket, ketalase; Pol, polymerase; Pol?, possible involvement with polymerisation or export; Exp, post-polymerisation processing or export.

DNA (Vanderslice *et al.*, 1989). The function of most of the gene products has been identified; the *Bam*HI restriction map in Fig. 7.2 indicates the order and approximate size of the fragments produced by the restriction enzyme. All seven transferase activities needed for the production of a fully acylated repeat unit were identified, as was a polymerase and three enzymes possibly involved in later stages of the synthetic process. Some properties of the different genotypes are shown in Table 7.3. Contrary to what might have been expected, the strains defective in adding either the side-chain terminal β-mannosyl residue or the terminal disaccharide (β-mannosyl–β-glucuronic acid) yielded polymeric products designated 'polytetramer' and 'polytrimer' respectively. This explained the lack of lipid intermediate accumulation shown in Table 7.3, and also indicated that the later stages of polymerisation were of relatively low specificity.

Mutants producing non-acylated and non-ketalated xanthan were also obtained. Through the use of recombinant DNA technology, strains with multiple defects were also constructed. These formed non-acetylated trimer or non-acetylated tetramer. The yield of polytetramer was 50% of that from wild-type bacteria; that of polytrimer was only 1–3%.

As well as the cluster of genes related to the synthesis of the repeat unit of xanthan, unlinked genes were found, concerned with the synthesis of the sugar nucleotides needed for xanthan production. One gene of the main cluster (*gumJ*) provided a lethal mutant, although lethality could be suppressed by mutations blocking initiation of xanthan synthesis: UDP–glucose pyrophosphorylase⁻ or glucose I transferase⁻. It was inferred that this mutant was unable to export the polysaccharide and that its accumulation in the cell proved to be lethal.

Rhizobium *species*

Polysaccharide synthesis in *Rhizobium meliloti* has been studied because of its relationship to effective nodule formation and symbiosis in leguminous

Table 7.3. *The functions of the* X. campestris *genes controlling xanthan synthesis*

Enzyme activity	Gene	Polymer synthesis	Lipid fraction
Glc transferase I	D	0	no sugar
Glc transferase II	M	0	glucose
Mannose transferase I	H	0	cellobiose
GlcA transferase	K	+	trimer
Mannose transferase II	I	+	none detected
Polymerase	E	0	pentasaccharide
Wild type		+	none detected

plants. Biochemically, the process resembles the synthesis of other extracellular heteropolysaccharides such as those of *E. coli*, *K. aerogenes* or *X. campestris*. However, the genetics and regulation of polysaccharide synthesis show significant differences from other bacteria. The products of a number of genes (*exo A–C, F, L, M, P, exoQ*) carried on a megaplasmid are involved in polysaccharide synthesis. Mutants in the *exoH* locus form non-succinylated polysaccharide (see p.31); *exoD, exoN* and *exoK* mutants produce lowered polysaccharide yields compared with wild-type bacteria. The products of two unlinked **chromosomal** loci, *exoR* and *exoS*, are involved in the regulation of exopolysaccharide synthesis. Both are negative regulators.

Some of the *R. meliloti* mutants are defective not only in the synthesis of succinoglycan, but also in the production of β-D-glucan and lipopolysaccharide; the majority of mutants are only defective in succinoglycan production (Table 7.4). The *exoB* and *exoC* mutations may cause defects in early non-specific steps common to the synthesis of all three carbohydrate-containing products. The *exoH* mutant, as well as failing to add succinyl groups to the exopolysaccharide, lacked any low-molecular-mass succinoglycan, although this is normally formed in wild-type strains.

Although little is yet known about export systems for exopolysaccharides, the *ndvA* gene product in *R. meliloti* appears to be involved in the export of an extracellular β-1,2-D-glucan. The gene coded a 67.1 kDa protein with homology to a number of bacterial ATP-binding transport proteins. An apparently equivalent gene, *chvA*, in *A. tumefaciens* also appears to function similarly and the gene product shows amino acid sequence homology to prokaryotic and eukaryotic export proteins. Evidence for complex regulatory systems for exopolysaccharide production in this bacterium has also been reported.

Table 7.4. *Polysaccharide synthesis in* Rhizobium *mutants*

Mutant	Succinoglycan	LPS	Cyclic β-glucan
exoA	−	+	+
exoB	−	altered	
exoC	−	altered	−
exoD	variable	+	+
exoF	−	+	+
exoG	−ᵃ	+	+
exoH	non-succinylated EPS	+	+
exoJ	trace	+	+
exoK	trace	+	+
exoL	−	+	+
exoM	−		
exoN	trace		
exoP	−		
exoQ	−		

Notes:
Genes *exo*R and *exo*S are involved in regulation; *ndv*A is involved in cyclic glucan export.
ᵃ Produces non-polymerised repeat units.

Haemophilus influenzae

Several features of polysaccharide-producing strains of *Haemophilus influenzae* type b have not yet been observed in other bacteria producing exopolysaccharides and may indicate rather different regulatory systems. A region of the chromosome involved in expression of capsule contains a directly repeated duplication of a DNA segment of approximately 17 kb. Loci in each copy are involved in exopolysaccharide production. Gene dosage effects were observed in strains without such duplication. Other *H. influenzae* serotypes lack the duplication.

The chromosomal segments in *H. influenzae* that are directly repeated are separated by a bridge region. In this region, a gene coding for polysaccharide export has been identified. Loss of the *bexA* gene results in loss of capsule expression. Thus one copy of the 17 kb segment together with a minimum 1.3 kb bridge region are needed for polysaccharide synthesis and excretion. The bridge region codes for a protein of *c.* 25 kDa, which has considerable homology with several proteins involved in energy-dependent molecular transport. It is also present in other *H. influenzae* serotypes. Strains from which the functional bridge region have been deleted accumulate polysaccharide within the bacterial cell, where it interferes with normal growth and division. The gene product may be coupled to ATP hydrolysis necessary for polysaccharide export, playing a role similar to the

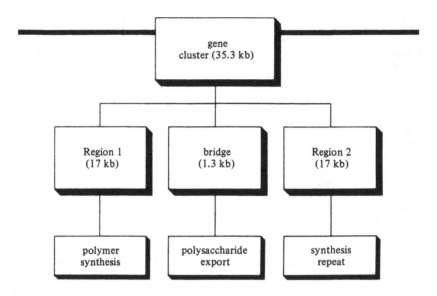

Fig. 7.3. The proposed genetic control mechanism for polysaccharide synthesis in *Haemophilus influenzae*.

inner membrane proteins in the transport **inwards** of maltose, histidine etc. (Kroll *et al.*, 1988). A high degree of homology has been detected between the polypeptide and the malK protein and others with ATP-binding sites.

Clearly, *H. influenzae* differs in the arrangement of chromosomal material from the cassette sequences found in enterobacterial systems (Fig. 7.3), but the requirement for an energy driven export system is likely to be a common feature of all exopolysaccharide-synthesising bacteria.

Acetobacter xylinum

The studies on xanthan have already permitted modification of polysaccharide structure both in respect of acylation and of carbohydrate side-chains. It should thus soon prove possible to modify other polymers composed of complex repeating units. It is necessary, however, to obtain considerable information about each polysaccharide-synthesising system to determine whether there any unique features. Bacterial cellulose synthesis in *Acetobacter xylinum* involves the formation of a low-molecular-mass, thermostable compound from GTP, which is an effective activator of cellulose synthase. The compound has been identified as an unusual cyclic guanyl nucleotide, composed of GMP residues, which is rapidly converted to 5′-GMP by *A. xylinum* membranes containing cellulose synthase activity. No comparable activators for other microbial exopolysaccharide-synthesising systems have yet been discovered.

A remarkable amount of information on the regulation of bacterial polysaccharide synthesis has become available in a very short time. It spans a range of different heteropolysaccharide structures but, as yet, little is known about homopolysaccharides or about synthesis in eukaryotes. However, it is clear from many of the prokaryotic systems which have been studied, that polysaccharide synthesis is frequently subject to very complex regulatory mechanisms.

Further reading

Boulnois, G. J. (1989). Genetics of capsular polysaccharide production in bacteria. *Symposium of the Society of Experimental Biology* **63**, 417–22.
Boulnois, G. J. & Jann, K. (1989). Bacterial polysaccharide capsule synthesis, export and evolution of structural diversity. *Molecular Microbiology* **3**, 1819–23.

8 Physical properties of exopolysaccharides

Studies of the physical properties of exopolysaccharides involve the application of a wide range of techniques; the interpretation of the results also requires a thorough knowledge of the chemical structure of the polymer. Electron microscopy has recently been applied to materials like xanthan to determine the persistence length of the molecule and to ascertain whether it is in a single- or double-stranded form. However, there are many limitations on interpreting the data thus obtained, owing to the possible introduction of artifacts both in the initial recovery of the polysaccharide and in sample preparation. Provided that the exact primary sequence and structure are known, X-ray fibre diffraction can supply information on polysaccharide conformation; circular dichroism provides a sensitive probe of the local environment of cation-binding sites.

8.1. Conformation

Determination of the molecular structure and conformation of a bacterial polysaccharide can be accomplished by X-ray diffraction of crystalline samples in the form of fibres. The techniques for orienting and crystallising the polymer use stress fields and annealing in the same way as they are applied to synthetic materials. The detail determined depends on the quality of the fibre diffraction pattern. The periodicity along the polysaccharide chain is visible in the approximately horizontal layer lines of the diffraction pattern. The spacing of the layer lines gives the pitch of the helical structure. Molecular model building can then be used, based on the known chemical repeating unit structure and standard values for bond angles and lengths and ring structures. Fig. 8.1 shows a typical X-ray fibre diffraction pattern for polysaccharide XM6 from *Enterobacter aerogenes*. It shows good agreement with the computer-generated projection of three-fold helical conformation.

In many of the polysaccharides which have been examined, the most favourable conformations display the charged portions of the macromolecules on the periphery, where they can readily interact with counterions and with water molecules. This may also be significant for *biological* recognition. On the other hand, neutral glucans such as mutan show stable conformations in which there is extensive intermolecular hydrogen bonding. This

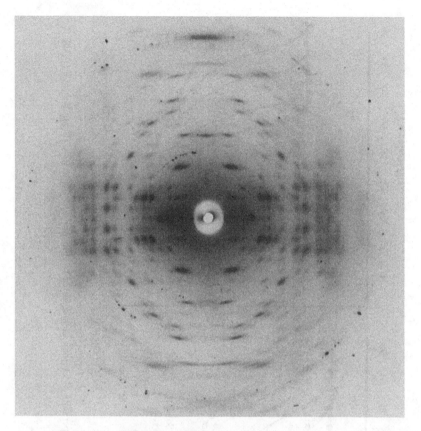

Fig. 8.1. X-ray diffraction pattern from oriented fibres of XM6
exopolysaccharide. The specimen was maintained at 97% relative humidity
during the exposure and exhibits a highly crystalline orthorhombic form (fibre
axis is vertical). The meridional periodicities are orders of 0.618 nm, less than
half the value of the fully extended trisaccharide backbone and is suggestive of a
double helical structure for the molecule. (Reproduced from Atkins *et al*.
(1986), by courtesy of Professor E. D. T. Atkins and the publishers.)

results in a sheet-like structure with alternating polarity of chain directions
within the sheet and consequent insolubility in water.

Other conformational information is obtained from studies of dilute
polysaccharide solutions. Light scattering and viscosity measurements
provide information on the ordered and disordered states of the scattering
and viscosity measurements provide information on the ordered and
disordered states of the polysaccharides and indicate the temperature at
which transition from one to the other occurs. Xanthan adopts an ordered
double helical conformation at lower temperatures; the presence of ions
influences this. The presence of salts raises the temperature at which the
conformational change is observed, as can be seen in Fig. 8.2 for two

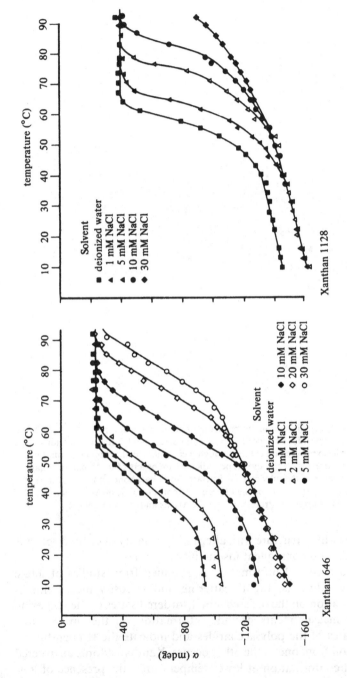

Fig. 8.2. Salt-induced changes in the transition temperature of xanthan as monitored by optical rotary dispersion. (From Shatwell (1989) with permission.)

xanthan preparations. Differences are seen in preparations with and without acetate and pyruvate groups; under some conditions, transition temperatures well in excess of 100 °C can be observed. The temperature of the transition mid-point can be reduced by removal of acetate or increased by removal of pyruvate. When pyruvate groups are present, they increase the internal repulsion between the charged side-chain termini. This may also be true of mutant xanthans defective in terminal mannose residues, leaving the β-glucuronosyl residues exposed.

The existence of 'families' of closely related polysaccharide structures has permitted an examination of the effect of relatively minor changes on the conformation of the polymers. Thus, xanthan lacking the terminal mannosyl residue from the side-chains has a similar ordered conformation to native xanthan. The pitch and symmetry are the same in both preparations, indicating that the presence of the terminal mannose is probably not essential for stabilising the ordered conformation of xanthan.

8.2.   Molecular mass

Determination of the molecular mass of microbial exopolysaccharides presents problems not associated with other molecular mass determinations. The polysaccharides in solution have a high viscosity, which presents difficulties in size exclusion chromatography. Until recently, accurate standards have not been readily available. Ultracentrifugation is also of limited use. The tendency for many polysaccharides to associate in solution has caused wide variations in the values recorded. It must also be remembered that the molecular mass *is* likely to vary, depending on the strains and cultural conditions used, and on the methods of isolation and purification (Table 8.1). The presence of microgels in polymers, such as xanthan, causes an increase in the apparent molecular mass; studies using quasi-elastic light scattering have demonstrated a time-dependent *rise* in the apparent molecular mass of xanthan. This results from aggregation of the polysaccharide molecules in deionised water. Morris *et al.* (1983), using light-scattering techniques, have shown similar effects when xanthan was heated in deionised water. The resultant relative molecular mass is 47×10^6, compared with 1.1×10^6 when heating was performed in 4 M urea, which breaks intermolecular hydrogen bonds and prevents reassociation of the macromolecules. Some examples of values for the relative molecular mass of xanthan, obtained by using different procedures, are indicated in Table 8.2.

8.3.   Emulsification

Many microbial species produce materials with surfactant activity during growth. However, only a small number of these have been extensively

Table 8.1. *Relative molecular masses*
of exopolysaccharides

Polysaccharide	M_r
alginate	5×10^5
dextran	$1 \times 10^5 – 2 \times 10^7$
elsinan	3×10^5
emulsan	9.9×10^5
pendulan	$1.6 \times 10^5 – 2.1 \times 10^5$
pullulan	$1 \times 10^4 – 1 \times 10^5$
scleroglucan	$1.9 \times 10^4 – 2.5 \times 10^4$
xanthan	$2 \times 10^6 – 5 \times 10^7$

studied. Some, such as *emulsan*, are capable of emulsifying hydrocarbon-in-water stable emulsions containing aromatic or aliphatic hydrocarbons. The bacterial product binds tightly to the surface of hydrocarbon droplets.

8.4.   Interaction with cations

As many of the microbial polysaccharides of potential industrial interest possess carboxylic acid residues from either uronic acids or pyruvate ketals, they readily complex cations. In some preparations, the interaction is particularly marked with ions of a specific ionic radius, but there is still interaction with other cations if they are present. Some polysaccharides form gels in the presence of certain cations, and the interactions can be followed without too much difficulty. Other polysaccharides do not produce such noticeable effects and various techniques have been used to study polysaccharide–cation interactions. These include circular dichroism, NMR spectroscopy, thermodynamic studies (microcalorimetry) and UV spectroscopy. The UV absorption spectra of the carboxylate groups are modifed considerably in the interaction with cations as polysaccharide–cation complexes are formed. Circular dichroism (CD) provides a particularly useful measure of interaction with univalent ions. CD ellipticity of polyguluronic acid is enhanced in the presence of K^+ and to a lesser extent Li^+, Na^+ or Rb^+. The spectral changes of poly-D-mannuronic acid are much smaller. In intact alginates, the size of the CD change correlates well with the poly-L-guluronic acid content.

The presence of different cations can markedly affect the conformation of the polysaccharide. In the sodium form, xanthan is a stretched coil, whereas the calcium form is a five-fold helix in which the divalent cations are strongly bound to the polysaccharide. Alginates interact very strongly with some divalent cations and are more selective in their binding than other polyuronides such as pectin. In studies with alginate, considerable

Table 8.2(a). *Relative molecular mass of xanthan*

Method	Value obtained
contour length (electron microscopy)	4×10^6 to 120×10^6
diffusion and sedimentation	1.4×10^6
intrinsic viscosity	62×10^6
light scattering	2.9×10^6 to 7.0×10^6 (extremes of 1.1×10^6 and 47×10^6)
low angle light scattering	1.5×10^6 to 5.2×10^6
sedimentation coefficient	14.8×10^6
translational diffusion coefficient (and intrinsic viscosity)	2.16×10^6

Note:
Many of these results represent reports from various authors. Light scattering has been the most widely used method.

Table 8.2(b). *Relative molecular mass of xanthans, obtained by static light scattering measurements*

Strain	$10^{-6} \times$ Relative molecular mass
X.campestris pv. *campestris* 646	0.9–1.2
X.campestris pv. *phaseoli* 1128	1.27
X.campestris pv. *phaseoli* 556	1.48
X.campestris pv. *oryzae* PXO61	1.60

Source: Results of Shatwell *et al.* (1990).

differences have been found between the interaction with Cu^{2+} and that with other ions. In contrast to Ca^{2+} there is little selectivity of binding, although some formation of 'junction zones' resembling those with Ca^{2+} and Sr^{2+} does occur.

Some microbial polysaccharides, including those from *Zoogloea* species, appear to bind ions very strongly and thus play an important role in flocculation. However, the exact mechanism for cation binding by this group of polymers has not yet been elucidated.

8.5. Rheological properties

Microbial polysaccharides vary considerably in their physical properties, including rheological characteristics. This has marked effects during fermentation. Many of the polymers show pseudoplastic flow (shear thinning), with apparent decrease in viscosity as the shear rate increases. This is seen for pullulan and xanthan as well as other polysaccharides Xanthan also displays a yield stress: a specific stress must be applied before movement of the fluid occurs. Xanthan solutions also have a tendency to regain their original state after release of stress, i.e. they demonstrate the combination of viscous and elastic behaviour known as viscoelasticity.

However, under conditions of low ionic strength, xanthan viscosity recovers less quickly after deformation. Nevertheless, the combination of rheological properties exhibited by xanthan solutions make them extremely useful for a wide range of industrial applications.

The fungal β-D-glucans such as scleroglucan and pendulan (from *Porodisculus pendulus*) both provide high viscosity in aqueous solutions. The viscosity is stable over a wide range of temperatures and, as these are neutral polymers, is unaffected by monovalent or divalent cations. It is also independent of pH in the range 2–12. The solutions resemble xanthan in being thixotropic. Typical random coil electrolytes in solution show large effects of pH and ionic strength on viscosity. Salt causes a fall in viscosity due to shielding of the charge and reduction in the random coil dimensions of the polymer. Because of the rigidity of the xanthan molecule, salt and pH have only minor effects on viscosity, except in very dilute polysaccharide solutions.

The rheological properties of a polysaccharide can be greatly affected by relatively small changes in chemical structure. Thus, curdlan is unbranched and forms gels, whereas scleroglucan, with the same main-chain structure but carrying β-D-glucosyl side-chains, forms highly viscous aqueous solutions. The effect of van der Waals interactions between the side-chain and main-chain portions of polysaccharide molecules has also been investigated in groups of polymers possessing closely related chemical structures. The gellan group are probably among the most suitable for such studies, as gellan is unbranched whereas the other members of this group all carry either single sugar or disaccharide side-chains. Significant differences in the physical properties of this group of microbial polysaccharides do not appear to be due to differences in the random coil conformations resulting from van der Waals interactions between side-chain and backbone. Interchain interactions are probably important and hydrogen bonding could also be involved.

8.6.   Gels

One of the most important properties of several of the industrially useful microbial polysaccharides is their ability to form gels. Some of these polymers require the presence of ions for gel formation (e.g. alginates or gellan) while others form gels without the involvement of ions (e.g.

curdlan). Gels owe their characteristics to the formation of networks of polymer chains which are cross linked. The actual mechanism of gelation varies; in some systems hydrogen bonding is involved, whereas in others there is covalent linking through multivalent cations. The site-binding of cations is important in a number of polysaccharides in maintaining the stability of the ordered conformation. Counter-ion selectivity appears to be explained by the sandwiching of arrays of cations, either between two folded buckled ribbons as in alginate or pectin, or between double helices as in carrageenan.

Curdlan (p. 21) is not soluble in water, but when an aqueous suspension is heated, it becomes clear at about 54 °C. Further heating leads to gel formation, as does dialysis against water of a solution in alkali. The gels are stable over the pH range 3–9.5 and do not melt at temperatures below 100°C. The gradual reduction in alkali concentration effected during dialysis permits a conformational change from *disorder* at high NaOH concentrations to *order* at 0.2–0.24 M NaOH. Adoption of the ordered conformation leads to gelation. High resolution ^{13}C NMR studies indicate that, at alkali concentrations greater than 0.2 M, the sharp and well-resolved spectra are characteristic of fluctuating, disordered chains (Saito *et al.*, 1977). However, cross-linking involving conformational ordering of the exopolysaccharide molecules yields gels in which a triple-helical structure is adopted. The gels formed by curdlan are thus effectively aggregated triple-helical regions. In the fungal β-glucans, such as schizophyllan and scleroglucan (p. 21), aggregation is inhibited by the presence of the β-D-glucosyl side-chains and no gelation occurs. Curdlan gels are stable over a wide pH range and are stable to freezing and thawing. Increase in the temperature of heating causes a further conformational change which increases the strength of the gels.

Alginates form gels by a very different mechanism: the selective cooperative binding of divalent cations. X-ray diffraction studies reveal a three-fold ribbon-like conformation for polymannuronic acid and a buckled two-fold structure for polyguluronic acid (Atkins *et al.*, 1973). Both types of structure bind calcium ions, but homopolymeric polyguluronate sequences of 18–20 residues or more show a sigmoidal increase in binding, which promotes the formation of junction zones. Mixed sequences and polymannuronic acid segments may have only minor roles. This cooperative binding process has been confirmed by competitive ion binding studies and by the large change in circular dichroism observed on addition of Ca^{2+} to alginate solutions or to polyguluronic acid preparations (Fig. 8.3). The specific site binding is seen in circular dichroism studies: the negative trough derived from the acid monomer sequences is converted to a positive value on controlled addition of Ca^{2+}. The magnitude of the observed difference is correlated in alginates with the proportion of polyguluronic acid sequences. Morris *et al.* (1975, 1978) thus proposed that

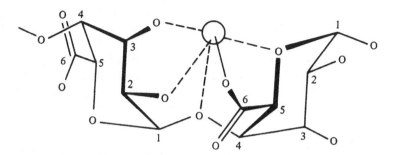

Fig. 8.3. Cation binding site of the dimer of L-guluronic acid in polyguluronic acid portions of alginate.

cross-linking of alginate gels involves the cooperative binding of Ca^{2+} in the interstices of aligned ribbons of polyguluronate sequences, giving a close ion-pair association with the carboxylate anion. This yields a so-called 'egg-box' model, which is relatively undisturbed by the removal of water (Fig. 8.4). The junction zones are terminated when the polyguluronate sequence gives way to polymannuronic acid or mixed sequences, which have a lesser role in gelation. Like curdlan gels, the alginate gels are thermostable and only melt at temperatures in excess of 120 °C. Clearly, only the microbial alginates that possess extensive polyguluronate sequences (i.e. those from *A. vinelandii*) (Fig. 3.5) can behave in a manner analogous to the alginates from the Phaeophyceae (brown algae).

The physical properties of alginate gels greatly depend on the ratio of D-mannuronic acid to L-guluronic acid. Brittle gels are obtained from alginate with a high guluronic acid content; weaker, more flexible gels result if the D-mannuronic acid content is high.

Gellan forms gels in the presence of either monovalent or divalent cations. The setting temperature and the strength of the gel depend on the nature and concentration of the cation as well as on the polysaccharide concentration. The gel contains polysaccharide chains organised in parallel to form an intertwined double helix. The molecules are half-staggered and each polysaccharide chain forms a left-handed three-fold helix with a pitch of 5.65 μm. In the gel formed with K^+, the cations are coordinated to the carboxylate group which is involved in interchain hydrogen bonding to yield a stable duplex. Gelation may even occur in two steps: chain ordering followed by chain association. In the reverse direction (heating), disaggregation occurs first, followed by melting of the individual helices. The gels are characterised by very marked hysteresis: the melting temperature is 60 °C and the setting temperature is 45 °C. In the absence of salt, the transition of gellan shows no detectable thermal hysteresis when studied by differential scanning calorimetry, i.e. there is a simple conformational transition from a double helix at low temperature to a single-coil structure

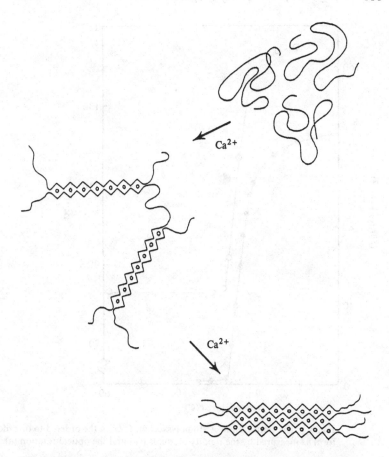

Fig. 8.4. The 'egg-box' model of alginate junction zones. Cross-links involve lateral association of sequences of L-guluronic acid to yield ordered junction zones, the Ca^{2+} ions occupying electronegative cavities in the structure.

at high temperature. Gels formed with K^+ set at lower temperatures than those with Ca^{2+}. Although gels of the same strength can be formed with K^+ or Ca^{2+}, the concentration of divalent cation needed is much lower. This is similar to the gelation of polysaccharide XM6 with Na^+ and Ca^{2+}, in which higher concentrations of monovalent ions are also required; the gel formed shows a very sharp transition at 30 °C, which is independent of the ion used (Fig. 8.5). Gellan differs from the polyuronides alginate and pectin, in the lack of specificity towards the ion bound. Monovalent ions induce gels in which the strength increases with atomic number. There is no selectivity within the alkali metal or alkaline earth cations; Mg^{2+} or Ca^{2+} have almost the same effect. The transition elements are more potent gel formers than the alkali earth ions and gel strength increased in the order $Zn^{2+} < Cu^{2+} < Pb^{2+}$. In welan gum, one of the gellan series (see Chapter

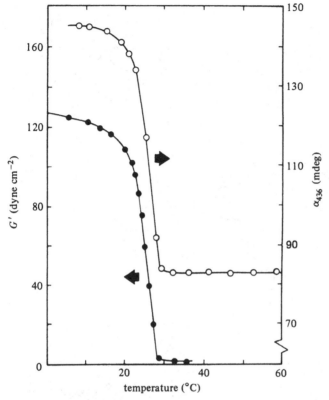

Fig. 8.5. The transition of XM6 polysaccharide from the ordered to disordered
form as measured by the rigidity modulus (G') and the optical rotation (a).

3), the presence of a side-chain stabilises the ordered structure but also
inhibits the aggregation process, which would be necessary for gelation.

The gel is in effect a metastable system, intermediate between the solid
state and the true solution. In the solid state, the molecules are in regular
ordered conformations, packed together in such a way that there is
relatively little hydration. In solution, the polysaccharide molecules are in
random conformations and are extensively hydrated. The integrity of the
gel structures is maintained through local intermolecular associations into
structurally regular *junction zones*. In these, the polysaccharide molecules
adopt the same ordered conformation found in the solid state. As a result,
there is a high degree of crystallinity in the junction zones, which are held in
association through regular non-covalent intermolecular bonds. Together
with the junction zones, segments of the polysaccharide molecules exist
which are structurally incapable of yielding stable interactions. These are
more highly hydrated than the junction zones and thus solubilise the
network of macromolecules. In their absence, no gelation would occur and
the net result of mixing would be either a solution or a precipitate.

The primary structure of a microbial exopolysaccharide has a very great effect on its capacity to form gels. This can be seen in a number of polysaccharides which possess closely related structures. Thus *curdlan* forms gels, whereas the structurally similar *scleroglucan*, differing only in the presence of glucosyl side-chains (Fig. 3.1), does not. Similarly, *gellan* is capable of gel-formation, whereas none of the other five polymers that are structurally related to it (Fig. 3.7) possess the ability to form gels although all are capable of forming highly viscous solutions. Very minor structural differences can affect the ability of polysaccharides to form gels. For example, the XM6 polymer readily forms gels with a number of ions, including Na^+ and Ca^{2+}, gelation being favoured when the interacting ion has an ionic radius of about 1.0. However, *Klebsiella aerogenes* type 54 polysaccharide, which differs from XM6 only in the presence of an *O*-acetyl group on *alternate* fucosyl residues of the tetrasaccharide repeat unit, does not form gels. Removal of the acylation (by mild alkali treatment) converts the K54 polymer into a gel-forming polysaccharide indistinguishable from XM6. Similarly, the native polysaccharide from *Pseudomonas elodea* is both acetylated and carries glycerol residues. It yields highly viscous aqueous solutions but it fails to form gels. On deacylation, the gel-forming commercial product *gellan* is obtained. In contrast to the native polymer, this material readily forms strong gels at relatively low polysaccharide concentration. It is obvious that, in each of these polysaccharides, the presence of ester-linked substituents reduces the structural crystallinity in the salt form and prevents gelation. The difference in crystallinity between the acetylated and non-acetylated *Klebsiella* polysaccharides has been clearly demonstrated by using X-ray fibre diffraction.

Synergistic gelling

Solutions of exopolysaccharides such as xanthan do not in themselves form gels, despite the strong intermolecular interactions which occur in solution. Some of the rheological properties of xanthan have, however, been explained on the basis of non-covalent association of the polysaccharide molecules to form three-dimensional 'weak gel' networks. Plant gluco- and galactomannans also normally fail to form gels, although gelation can be induced by certain ions or under reduced water activity. However, when solutions of xanthan and gluco- or galactomannans are mixed, heated and cooled, firm gels may be formed at relatively low polysaccharide concentrations. This is also true of a small number of other polysaccharide mixtures, but our understanding of the process comes mainly from studies on the xanthan–galactomannan system. The interaction between the bacterial and plant polysaccharides has been identified on the basis of chiroptical, rheological and NMR relaxation evidence as an energetically favourable association of structurally and sterically regular portions of the polysaccharide chains (Dea *et al.*, 1977). Although synergism of this type is seen when xanthan, which retains its rigid ordered structure in solution, is mixed with plant galactomannans, no such interaction is observed with bacterial polymers forming random coils. Synergistic gelling is obtained when the mixtures in solution are heated above the helix–coil transition temperature (95 °C) of xanthan, then cooled to ambient temperature. No gelation occurs when sufficient Ca^{2+} is present to raise the transition temperature above 100 °C. Thus denaturation of the xanthan helix must occur if intermolecular association and resultant gelation are to be found.

Interactions between different xanthan preparations and a wide range of plant gluco- and galactomannans have been investigated. In general, the most effective galactomannans in synergistic gelling reactions with xanthan are those that possess mannan backbones with low galactosyl substitution. It can be shown by enzymic removal of the D-galactosyl residues that the degree of interaction between samples of a galactomannan, such as gum guar and xanthan, increases as the galactose content decreases. Thus, it appears likely that the interaction between the two types of polysaccharide molecule primarily involves unsubstituted regions of the mannan chains and only small segments of each molecular species participate. Therefore, the greatest interaction is found with those plant galactomannans in which there are regions lacking the galactosyl side-chains; similar polysaccharides in which there is a comparable galactose content but in which these residues are evenly distributed along the chains show only weak interaction with xanthan. Examples of this effect are seen with the galactomannans from *Ceratonia siliqua* (locust bean gum) and *Caesalpinia pulcherrima*, each of which contains 24–25% of galactose. In the former, the galactosyl residues are **not** randomly distributed whereas in the *C. pulcherrima* polysaccharide the distribution of galactose is statistically close to random. From this, it is clear that only locust bean gum has blocks of mannose residues relatively free of galactose and available for interaction with xanthan. One apparently anomalous result was the observation that xanthan interacts strongly with the polysaccharide of high galactose content from *Leucaena leucocephala*. However, enzyme studies have indicated that the galactose side-chains in this polymer are principally arranged on *alternate* backbone residues and are consequently all on the *same* side of the macromolecule. The bacterial polysaccharide thus interacts with the unsubstituted side of the molecule rather than with sections of the plant polysaccharide that lack side-chain sugars (McCleary *et al.*, 1985; Dea *et al.*, 1986). An alternative explanation of the process has been offered by other workers who suggest a 'lock and key' arrangement between the xanthan side-chains and the galactomannan backbone.

In addition, xanthan interacts strongly with glucomannans such as the partly acetylated polymer from *Amorphophallus konjac* (konjac mannan). The strength of this interaction can be seen by comparing the melting point of a mixed gel of xanthan and konjac mannan with the values obtained for xanthan–galactomannan mixtures. The highest values found for the latter are 41 °C, whereas a xanthan–konjac mannan mix of 4 : 5 with 0.45% total polysaccharide melts at 63 °C. Gels are also formed at total polysaccharide concentrations as low as 0.02–0.05%. Xanthan does show interaction with other $(1\rightarrow4)$-β-linked polysaccharides and their derivatives. Thus mixtures of xanthan and either hydroxyethylcellulose or carboxymethylcellulose show enchanced viscosity. The phenomenon of synergistic gelling is widely used in industry and has numerous food applications (Chapter 9), including in particular canned pet foods.

A galactomannan from *Leucaena leucocephala* contains 40% galactose but is also capable of strong interactions with xanthan. It has regions of high galactose substitution and in the two-fold ribbon conformation adopted in the ordered state, unsubstituted 'faces' are available for interaction. The distribution of the galactosyl side-chains of the galacto-mannans therefore greatly affects the synergistic gelation.

The fine structure of the xanthan molecule also plays a part. If the xanthan molecule is heavily acetylated, higher concentrations are required compared with non- or poorly acetylated polysaccharide. Clearly, the strengths of the interactions between the two types of polysaccharide are influenced by the side-chain components of both molecules. Of the commonly used galactomannans, locust bean gum (LBG), with a mannose to galactose ratio of *ca.* 4 : 1, gels with almost all the xanthan preparations tested. Guar gum (mannose : galactose, 2 : 1) only forms gels with a very small number of xanthan preparations and appears to be greatly influenced by the presence of acyl groups (Table 8.3). The mixtures form true gels, which recover neither from mechanical damage nor from flow. They are thermally reversible with a sharp melting point and setting point within a 2–3 K temperature range. The actual values increase with *total* polysaccharide concentration being essentially independent of the *ratio* of the two polysaccharides, when mixtures of 1 : 1, 1 : 2 and 1 : 3 xanthan : galacto-mannan are tested. The melting temperature at 0.5% total polysaccharide concentration is 31–32 °C, rising to 42 °C at 2% polymer concentration. In the mixed gels, the ordered conformation of xanthan persists. It has been suggested that xanthan–galactomannan gels may be cross-linked by cooperative association of non-substituted regions of the mannan in a regular ribbon-like conformation, with the ordered xanthan structure. Increase in the galactosyl substitution of the galactomannan or in the acetylation of the xanthan reduces the interactions and consequently the extent of gelation. Salts also affect the interactions, weakening the gels or preventing their formation.

The wide range of chemical structures found in microbial polysaccharides yield an extensive spectrum of physical properties. Some of these permit the industrial application of the polymers, mainly as viscosifying or

Table 8.3. *Gelling interaction between xanthans and gluco- and galactomannans*

Each set of results represents a different xanthan preparation before and after chemical deacylation.

Xanthan			Lowest dilution for gelation	
Acetate	Pyruvate[a]	Gluco/Galactomannan	Xanthan	Mannan
1	0.4	LBG	0.063%	0.125%
		guar	0.25%	0.5%
		konjac	0.25%	0.5%
1	0	LBG	0.125%	0.25%
		guar	no gelation	
		konjac	0.5%	1.0%
2	0	LBG	0.125%	0.25%
		guar	no gelation	
		konjac	0.25%	0.5%
0	0	LBG	0.063%	0.125%
		guar	0.25%	0.5%
1.1	0	LBG	0.25%	0.5%
		guar, konjac	no gelation	
0	0	LBG	0.031%	0.063%
		guar	0.5%	1.0%
		konjac	0.125%	0.25%

Note:
[a] Molar ratio.

gelling agents. In many cases, the full extent of the physical properties is only now being determined and it is not yet possible to match these with the requirements of specific applications.

Further reading

Brant, D. A. (1981). *Solution properties of polysaccharides.* ACS Symposium Series no. **150**. American Chemical Society, Washington.

9 Food usage of exopolysaccharides

Polysaccharides are incorporated into foods to alter the rheological properties of the water present and thus change the texture of the product. Most of the polysaccharides used are employed because of their ability to thicken or to cause gel formation (Table 9.1). Advantage is also taken of the ability of some mixtures of polysaccharides to exhibit synergistic gelling: basically, for the two polymers to yield a gel at concentrations of each which will not in themselves form gels (Chapter 8). Associated with these readily measurable properties are others, such as 'mouth feel', which are more difficult to define but which also show some correlation with physical properties. 'Mouth feel' has been related to viscosity and, in particular, to non-Newtonian behaviour. This also relates to the masking effect of viscosity on the intensity of taste. There is also a specific relationship between the polysaccharide and flavours present in any food. Thus, corn starch and xanthan both provide good perception of sweetness and flavour when compared with gum guar or carboxymethylcelluose. In addition, polysaccharides are used because of their capacity to control the texture of foods and to prevent or reduce ice crystal formation in frozen foods; they may also influence the appearance and colour as well as the flavour of prepared foodstuffs (Table 9.2). It must also be remembered that many foodstuffs already naturally contain polysaccharides such as starch or pectin. Thus, addition of any further polysaccharide or polysaccharides will in all probability involve interactions with these compounds as well as with proteins, lipids and other food components.

The use of polysaccharides in food is also governed by a number of factors unrelated to their physical properties. Currently there is considerable consumer concern over the use of 'food additives' and these include hydrocolloids. Most regulatory bodies review their lists of permitted food additives at regular intervals, when they may amend the regulations. This may also lead to improved or altered specifications.

Different customs exist in different countries. Unlike North America and Europe, in Japan microbial polysaccharides are regarded as *natural* products, which can be added to food without specific regulatory control. There is thus an upward trend in the use of microbial exopolysaccharides such as curdlan in Japan. There may be market shortfalls in some of the *natural* (plant or algal) polymers. This can be exemplified by the fall in the supply of gum arabic in 1985 to 25% of the 1984 level. Shortfalls have also

117

Table 9.1. *Polysaccharide properties used in food*

Function	Application
adhesive	icings and glazes
binding agent	pet foods
coating	confectionery
emulsifying agent	salad dressings
encapsulation	powdered flavours
film formation	protective coatings, sausage casings
fining (colloid precipitation)	wine and beer
foam stabiliser	beer
gelling agent	confectionery, milk-based desserts, jellies, pie and pastry fillings
inhibitor (crystal formation)	frozen foods, pastilles and sugar syrups
stabiliser	ice cream, salad dressings
swelling agent	processed meat products
syneresis inhibitor	cheeses, frozen foods
synergistic gel formation	synthetic meat gels, etc.
thickening agent	jams, sauces, syrups and pie fillings

been seen recently in the supplies of algal alginate, locust bean gum and gum guar. The drought in the Sahel region alone is thought to have reduced the supply of plant exudate and other gums from the area by 26 000 tonnes. These shortages in plant or algal gums may lead to increased use of microbial products in their place and provide an opportunity for the introduction of new microbial products.

9.1.   Natural occurrence of microbial polysaccharides in foods

As well as the deliberate addition of microbial polysaccharides to food products to obtain specific properties, there are a number of bacterial fermentations in which polysaccharide is produced and is needed to yield a specific type of product. An example of this can be found in certain types of fermented milk product such as yoghourts. In some of these, the production of polysaccharide during bacterial growth is claimed to enhance the product, particularly in respect of the body and texture of the product and in its smoothness and mouth feel. This is particularly true in countries such as the Netherlands and France, in which the addition of plant or animal stabilisers is prohibited. The polysaccharide from *Streptococcus salivarius* subsp. *thermophilus* is an example of this. Unfortunately, the production of an apparently neutral glycan by these bacteria is unstable and the use of the bacteria tends to lead to a lack of uniformity in the product. Polysaccharide-producing strains of *Lactobacillus bulgaricus* are also used for this purpose, strains being available which synthesise from 14 to 400 mg, per litre of culture, of a viscous polysaccharide in which galactose is the major

Table 9.2. *Factors in polysaccharide choice for food use*

1. Type of application
2. Viscosity or gelation
3. Mouth feel
4. Appearance (including clarity, etc.)
5. Emulsification capabilities
6. Compatibility with other food ingredients
7. Synergistic or other effects
8. Stability to physical and biological factors under the conditions of intended usage
9. Cost
10. Acceptability (legal, consumer and otherwise)
11. Quality: odourless, tasteless, consistent quality, etc.

component. There are also some fermented milk products in which the production of a thick, gel-like texture results from the exopolysaccharide synthesised by the bacteria used. These are traditional products from parts of Finland, termed 'villi', which do not have widespread sales elsewhere.

9.2.   Emulsions

A wide range of foods are essentially colloids in which there are complex interactions between the various ingredients. Emulsions of oil-in-water, or water-in-oil, form an important group of food products in which oils or lipids, water and other ingredients are processed to form sauces, spreads, etc. Some of the emulsion-containing systems are dried and then reconstituted, as in packeted instant soups, desserts and sauces. Whether used after the initial processing or after reconstitution with water, such emulsions may require to be stabilised with polysaccharides. In this role, xanthan has found many applications in dry-mix food products. It can be dispersed in either cold or hot water to provide thixotropic dispersions which can be subsequently heated or refrigerated. It prevents the constituents from reverting to their original separate phases and ensures the long-term stability required for food products.

Some of the foodstuffs in which polysaccharides are incorporated are of relatively low pH. In this category are salad dressings, relishes and yoghourts. Thus any polysaccharide incorporated into these foods must be acid-stable. Xanthan, with its stability over a wide pH range, is well suited to such applications.

9.3.   Gelling agents

A widespread usage of polysaccharides in the food industry is to gel the aqueous phase. Various types of interaction may produce the gel, including hydrogen bonding and ionic interactions. These interactions on their own are relatively weak, but when they occur in large numbers, ordered polysaccharide conformation is produced. Also, the single polysaccharide chain can enter into several regions of ordered structure, through either intermolecular or intramolecular interactions. The result is a three-dimensional network, i.e. the gel structure is formed. Termination of the region of ordered structure occurs when the sequence of monosaccharides in a polymer chain alters, thus preventing further association. The final property of the gels is determined by the relative proportion of polysaccharide chains that are involved in ordered associations. This can be exemplified by alginate gels, in which the major role in gel formation is played by the poly*guluronic acid* sequences. When these terminate, being replaced by either mixed sequences or poly*mannuronic acid* blocks, the association brought about by calcium ions ceases. This also demonstrates why *A. vinelandii* alginate, alone of the bacterial alginates, behaves in the same manner as the products from marine algae. Alginate gels have the advantage that they permit re-forming of food products into shapes that are not altered by subsequent heating. Thus homogenates of pimento can be converted, after mixing with alginate and Ca^{2+}, into sheets of uniform texture and thickness, which are then cut and inserted into cocktail olives. Similarly, a range of meat products, which hold together either in the chilled state or when frozen, can be prepared. An advantage obtained when using alginate as a binder is the elimination of the need for added salt and polyphosphates.

The gelation of foodstuffs employs various polysaccharides, starch being very widely used because of its low cost and ready availability and acceptability. However, it is not suitable for all food applications. Indeed, although a wide range of polysaccharides can be used as gelling agents for foods, they are not always interchangeable, as each has specific gelling properties. Agar, carrageenan and xanthan–galactomannan mixtures form thermoreversible gels. Gelation occurs on cooling the mix and can be reversed by heating. Alginate and pectin with low methoxyl content are different. They react with Ca^{2+} to yield non-thermoreversible gels.

In some applications, it does not matter whether the gel used is clear or not. Thus, in meat or milk gels, any opacity in the gel does not matter. However, clear gels are needed for many purposes. Gelatine is employed widely for this purpose at present, but agar and alginates also find a number of applications. Once it has official approval, the very clear gels formed by low levels of *gellan* can be expected to have considerable aesthetic as well as practical appeal. The potential usefulness of gellan as a gelling agent for the food industry in part derives from its similarity in physical properties to carrageenan. Both require monovalent or divalent ions for gelation to proceed. Considerable hysteresis is observed on melting and gelling and the strength of the gel depends on the nature and the concentration of different

Table 9.3. *Potential food applications of gellan*

Dairy products	Ice cream, milkshakes, yoghourts
Fabricated foods	Fabricated fruits, meat etc.
Icings and frostings	Bakery icings, frostings
Jams and jellies	Low calorie jams, jellies, fillings
Pie fillings and puddings	Instant desserts, jellies
Water-based gels	Aspics
Pet foods	Gelled pet foods, meat chunks

cations present. Other hydrophilic food ingredients, such as sucrose, tend to reduce the required concentration of ions. Gelation occurs over a wide pH range with little significant effect on gel strength. The gels are very clear indeed and the level of polysaccharide needed is often below that for conventional hydrocolloids. The wide range of potential food applications for gellan as a gelling or texturing agent is shown in Table 9.3 (Sanderson and Clarke, 1983). The gels are stable to autoclaving and are in general, thermally reversible. The advantage in using gellan to replace other polysaccharides is that, in most preparations, the gum concentration is lower and the process may be performed (in some cases) more quickly. Thus, use in starch jellies for confectionary can halve the preparation time (12 h instead of 24–48 h) and only 0.07% gellan is needed. Jellies made with gellan are extremely clear; only 0.2% gellan is required to give a mouth feel similar to that of gelatine.

In Japan, curdlan has been shown to be of use for improving the texture of various foods, including noodles, sausage, jams and jellies, and soy-bean curd. Curdlan also retains shape in cooked foods and improves their thickness and stability when used at concentrations ranging from 0.05 to 3%. The amount of polysaccharide added depends on the food and the purpose for which the polymer is used. When incorporated into jellies, it has the advantage that these can then be prepared at relatively low temperatures (55–80 °C) and destruction of flavours is avoided. When dried gels are re-formed in water they retain their original shape. Curdlan has also been proposed as a binding agent for fishfoods.

Synergistic gelling of galactomannans and xanthan is also utilised in a number of foods, including cheese spreads, cream cheese and pie fillings (Table 9.4). In the cheese products, the mixture of polysaccharides permits easier slicing and also gives good mouth feel and flavour release. The polysaccharides are also included in beverages to control the sedimentation of particulate material. They are incorporated into ice-creams and similar freeze–thaw products to provide viscosity and other attributes.

Table 9.4. *Food usage of xanthan–galactomannan mixtures*

Gel formation and stabilisation	Ice cream
	Milk shakes, milk drinks
	Dessert gels
	Puddings and pie fillings
	Cheese and cream cheese
Viscosity control	Ice cream
	Milk shakes
	Chocolate drinks
	Instant soups
	Cottage cheese dressings

9.4.   Legislative acceptability

Before a new microbial polysaccharide can be permitted for use as a food additive, it must be submitted to a process of approval. Quite apart from the technological justification for the inclusion of the polymer in foodstuffs, evidence must be provided of the *need* to use it and of its safety in use. The producer must demonstrate that the new additive will benefit the consumer. This covers several categories:

(i)   presentation of the foodstuff;
(ii)   the need to keep the food wholesome until eaten;
(iii)   the extension of dietary choice;
(iv)   the need for nutritional supplement;
(v)   convenience of purchasing, packaging, storage, preparation and use; and
(vi)   economic advantage, including longer shelf-life or reduced price.

Clearly not all these factors will be appropriate in the case of microbial polysaccharides. Research must first be performed by the firm intending to produce or market the polysaccharide. Evidence in the UK is then presented to the **Food Advisory Committee**, in which there are representatives of consumer and enforcement bodies, the food industry, medicine and academia. This advises the relevant Government Departments (Health; Agriculture, Fisheries and Food). If the Food Advisory Committee considers that the *need* has been proved, safety is then examined (WHO, 1987). This is assessed by the **Committee on Toxicity of Chemicals in Food, Consumer Products and Environment.** The tests include determination of the LD_{50} in rats, feeding trials to animals through two generations of reproduction, and teratology testing. Tests for dust inhalation must also be performed. The probable costs for a development programme of this type may well be of the order of US$ 10 million and the time that must therefore elapse, before approval is given, may be from 3 to 6 years. In addition to official approval for the additive, there must be consumer acceptance. One major factor in favour of the microbial polysaccharides is their lack of digestion by humans and animals. They can thus be used as constituents of low calorie diets as well as other processed foodstuffs. Although the polysaccharides are not normally digested by animals, examples of degradation by anaerobic intestinal microorganisms have occasionally been reported.

Once the Food Advisory Committee has made its recommendation, it will still need approval from the EEC if the foodstuffs in which it will be incorporated will be marketed in other member countries of the European Community. In the UK, there is a schedule of 'Emulsifiers and Stabilisers in Food Regulations' promulgated by the Ministry of Agriculture, Fisheries and Food. The European Community has attempted to harmonise food additive regulations within member countries through a *Directive of the Council of Ministers* on 'Emulsifiers, Stabilisers, Thickeners and Gelling Agents for Food Use' (74/329/EEC). This provides a list of the agents that are fully accepted in member states. The accepted polysaccharides (and all other additives) are designated with a serial number, that for xanthan being E415. This is displayed on the packaging of foodstuffs in which it is incorporated.

The World Health Organisation (WHO) and Food and Agriculture Organisation (FAO) have formed a joint expert committee on food additives, which has the responsibility of proposing **acceptable daily intakes** (ADIs). The Committee divides food additives into three categories.

List A. Fit for use in food.
List B. Need to be evaluated.
List C. Should not be used in foods.

After evaluation, the food additives are detailed in specifications for identity and purity of some food additives.

In the USA, xanthan is permitted as a food additive under regulations controlled by the Food and Drugs Administration. It is on the list of substances 'Generally Regarded as Safe' (GRAS list), being approved in the Federal Register for use as a stabiliser, emulsifier, foam enhancer, thickening, suspending and bodying agent. Its use is permitted in sauces, gravies and coatings applied to meat and poultry products. It may also be added to cheeses and cheese products, certain milk and cream products, food dressings, table syrups and frozen desserts. A further use for xanthan, which also requires FDA approval, is as a component of paper and paperboard which is intended for use in food packaging and will possibly be in contact with the foodstuffs. The xanthan may be used in coatings of papers and boards. Although dextran received approval for food use and a number of potential applications have been suggested, there is apparently no current food usage of this polysaccharide. Curdlan, which might have a number of food applications, is used in Japan as a *natural* material, but is not permitted in the USA. The next microbial exopolysaccharide that appears likely to receive consideration for use as a food additive is gellan, the gelling properties of which could well find numerous applications in food preparations.

The definition of exopolysaccharides for legislative purposes and for patents is not particularly satisfactory. They are generally identified by the molar ratio or the composition of their carbohydrate and non-carbohydrate components. Occasionally, more detailed information on structure may be provided, together with properties such as optical rotation, viscosity, physical appearance and ash content. As more exopolysaccharides are proposed for food usage, it is probable that some will have similar composition but different structural details and that improved methods of definition will be needed. It is also surprising that few methods have been developed for the identification and quantification of the exopolysaccharides once present in food. With the discovery of enzymes specifically acting on the exopolysaccharides, these could well be used to provide highly specific assay procedures for the polymers once they have been incorporated into foods containing other carbohydrate-containing materials. Alternatively, techniques based on monoclonal antibodies could be used.

Xanthan is currently described in the National Formulary as:

> '. . a high molecular weight polysaccharide gum produced by a pure culture fermentation of a carbohydrate with *X. campestris*, then purified by recovery with isopropyl alcohol, dried and milled. It contains D-glucose and D-mannose as the dominant hexose units, along with D-glucuronic acid, and is prepared as the sodium, potassium or calcium salt. It yields not less than 4.2% and not more than 5% carbon dioxide, calculated on the dried basis, corresponding to not less than 91% and not more than 108% xanthan gum.'

Food-grade xanthan must also meet the specifications listed in the Food Chemicals Codex, which include limits on arsenic (< 3 ppm), heavy metals (< 30 ppm), isopropanol (< 750 ppm) and pyruvate ($< 1.5\%$). It must also be satisfactory with regard to the absence of *Salmonella* species and *E. coli*.

9.5.   Exopolysaccharides as a source of flavour components

As many microbial exopolysaccharides contain appreciable amounts of 6-deoxysugars, it has been suggested that the polymers might be used as sources of these sugars (Graber *et al.*, 1988). The 6-deoxyhexoses, in turn, could then be used as intermediates in the synthesis of furaneol and its derivatives. These compounds can be used as flavouring agents for the food industry. Furaneol yields a powerful caramel-like flavour, but when modified by the addition of various short-chain fatty acid esters, gives a range of either meat or fruit flavours. Furaneol is fairly expensive (*ca.* US$ 100–200 per kilogram); it has been proposed that this cost could be reduced through the use of polysaccharide-derived deoxysugars. If this were to be achieved economically, it would require release of the sugars from the polymers by enzymic treatment. By using immobilised enzymes or cells and polysaccharides containing a high percentage of deoxysugars, this might be feasible; however, enzymes capable of degrading the appropriate exopoly-saccharides to their component sugars have not yet been described.

Further reading

Blanshard, J. M. V. & Mitchell, J. R. (1979) *Polysaccharides in food.* Butterworths, London.

Mitchell, J. R. & Ledward, D. A. (1986). *Functional properties of food macromolecules.* Elsevier, Amsterdam.

Phillips, G. O., Wedlock, D. J. & Williams, P. A. (1986). *Gums and stabilizers for the food industry*, vol. 3. Elsevier, London. [Also other volumes in this series.]

Wheelock, V. (1986). *Food additives in perspective: A review of current issues with particular reference to consumers, industry and government.* University of Bradford.

Yalpani, M. (1987). *Industrial Polysaccharides.* Elsevier, Amsterdam.

10 Industrial uses of microbial polysaccharides

Both microbial and non-microbial polysaccharides find a wide range of non-food industrial uses. In such usage, the polysaccharides may compete with synthetic organic polymers, but in some applications only natural products are acceptable because of their biodegradability and their lack of toxicity. As far as microbial polymers are concerned, xanthan has the largest share of the market. The non-food usage of polysaccharides directly reflects their various physical attributes. Before their use is contemplated, they must be fully evaluated against other possible synthetic and natural products. If these other products are cheaper, they may be preferred even though the microbial polysaccharide is superior in the application envisaged. They are more expensive than starch or than synthetic products such as polyacrylamides, so use of the exopolysaccharide may incur an unacceptable cost penalty. However, because of changes in prices and availability of plant and algal products, there is considerable opportunity for expansion of the industrial use of microbial polysaccharides, especially as new products with unique physical properties are discovered. The oil industry provides one major scenario in which exopolysaccharides have readily found acceptance for a number of purposes, in competition with plant gums and their derivatives and with synthetic chemicals. If the numerous proposed enhanced oil-recovery developments come to fruition, the use of xanthan and other microbial exopolysaccharides could well increase very significantly indeed. Other industrial applications have secured a wide range of technological uses for exopolysaccharides with different physical properties. Some of these are limited in scale and will require only small amounts of polymer, but others contribute significantly to the overall demand.

10.1. Biopolymer use in the oil industry

Drilling

Rotary drilling uses a cutting head or bit, which is at the end of a rotating hollow pipe. The bit is of larger diameter than the pipe. Consequently, there is an annular space between the pipe and the rock formation in which the drilling is taking place. Fluid pumped down the pipe passes the drilling bit and returns to the surface via the annular space. This fluid serves the dual functions of lubricating agent and suspending fluid for the removal of rock cuttings from the drilling head and carrying them to the surface. The drilling **mud** or drilling fluid used, whether in the drilling of exploratory or of production wells, is carefully formulated to yield specific properties in respect of fluid flow and rheology. The fluid must be non-corrosive and must always be compatible with any salts which are used to protect the rock formation and minimise its interaction with the drilling fluids. These salts include lime, gypsum and potassium chloride. The drilling fluid must have pseudoplastic rheology and must be stable under the shear forces and temperatures encountered during drilling. It must contain biocides to prevent the destruction of the dissolved biopolymers and the growth of (anaerobic) bacteria which generate hydrogen sulphide and carbon dioxide. Polysaccharides have been used for some time in drilling mud formulation; xanthan in particular has found world-wide applications in drilling mud preparations and in the completion fluids employed to ensure free flow of oil. This reflects its exceptional compatibility with salts, suspending capacity, pseudoplasticity in solution and stability at high temperatures and extremes of pH. In some wells, succinoglycan with a lower transition temperature than xanthan can be advantageously used. Succinoglycan is also less susceptible to interaction with $CaBr_2$, which is used to provide density in drilling fluids. The polysaccharide solution has good carrying capacity and minimal salt impairment of viscosity. It is used in completion operations such as gravel packing, in which the gravel (coarse sand) is pumped down a well as a cold suspension to provide a mesh to maintain fluid flow. As the suspension heats up *in situ*, viscosity is irreversibly lost, a process aided by the addition of HCl to promote degradation, and the gravel packing is completed.

Enhanced oil recovery

Oil recovery is limited by many factors, including reservoir pressure and pressure differential, oil viscosity, reservoir formation structure and permeability, reservoir temperature, etc. The level of recoverable oil can be enhanced by maintaining the reservoir pressure through the re-injection of either produced or surface water, or of gaseous hydrocarbons or carbon dioxide. The water has to be filtered, freed from dissolved oxygen and suspended material and treated with biocides. Precipitation or deposition of solids must be avoided or the formation will be plugged and further injection of water prevented. In a typical large-scale operation, this is a very expensive practice in itself, requiring the processing of millions of gallons of either surface or connate water.

Further oil may be recovered through polymer flooding and/or surfactant flooding, both of which involve the use of large quantities of relatively expensive chemicals. The aim of the polymer flooding is to increase the viscosity of the water, thus increasing the sweep efficiency and

Table 10.1. *Properties of exopolysaccharides for use in enhanced oil recovery*

High viscosity in water and in concentrated salt solutions
High shear stability
Pseudoplastic
Stable over wide pH range
Stable for prolonged periods when exposed to high temperatures
Freedom from particulate material; high injectivity
Low adsorption to reservoir rocks

preventing the water from by-passing the oil. In surfactant flooding, surface tension is reduced and more oil can then be recovered. If the oil is less mobile than the displacing fluid (water), the water is likely to by-pass the oil. This results in poor sweep efficiency and little enhancement of oil recovery. The practice of dissolving polymers of high molecular mass in the injection water reduces its mobility and lessens the chance of its by-passing the oil, with resultant increases in the yields of oil recovered.

The limitation on the use of polymers for mobility control during water injection into oil reservoirs is the range of conditions encountered in such environments. Few are ideal for polymer use and each reservoir must be assessed separately. Both synthetic polymers (polyacrylamides or petroleum sulphonate) and microbial exopolysaccharides have properties which may render them suitable for application in enhanced oil recovery. A decision on which type of polymer to employ for water flooding must be taken in the light of the temperature, salinity, pH etc. Even high concentrations of synthetic polymers can prove ineffective under conditions of high salinity; at lower salinities, the use of synthetic compounds requires much higher concentrations to achieve the desired viscosity. If a biologically produced polymer such as xanthan or scleroglucan is used, the concentration needed is much lower and it is not affected in the same way by high salinity (Table 10.1). Further, temperature does not greatly affect the solution viscosity provided that there is no thermal degradation of the molecule (i.e. solutions are stable at temperatures of up to 90 °C). One of the key factors thought to be important in maintaining the structure of the biopolymers at high temperature, is to ensure that they are not subjected to free radical attack. In this, the presence of Fe^{2+}–Fe^{3+} complexes, together with small amounts of dissolved oxygen, plays an important role. The concentration of exopolysaccharides used for polymer flooding in enhanced oil recovery is limited by the relatively high cost of the biopolymers and by possible injectivity problems. Potential usage is probably limited to small- to medium-sized reservoirs, in which the oil viscosity ranges from 10 to 100 cP and the temperature does not exceed

Table 10.2. *Properties of UK offshore oil reservoirs*

Property	Forties	Brent	Upper Beryl	Auk	Thistle
depth (feet)	6800	8500	10000	7700	9300
temperature (°C)	90.5	93.3	97.7	—	102.3
oil viscosity (cP)	0.8	0.28	0.52	1.2	—
formation water salinity (p.p.m.)[a]	102000	—	—	—	—

Note:
[a] Total dissolved salts.
Dashes indicate that operators have not made figures public.

80–90 °C, although in the salinity of seawater xanthan has an estimated transition temperature of *ca.* 120 °C. The stability of the polymer can be still further increased in the presence of potassium acetate. The pressure of many of the deeper reservoirs does not reduce the transition temperature by more than 5–10 °C. When the reservoirs are shallow and consequently at moderate temperatures, biocides must be added to prevent the biological destruction of the polysaccharides. As the initial dilution and injection are at ambient temperature, the presence of biocides is essential if degradation of polymer is to be avoided.

Polymer flow in porous media is a very complex process, owing to the various physical phenomena which occur. A factor which has to be taken into consideration in deciding whether to use any polymer for enhanced oil recovery, is the adsorption of the molecules onto the reservoir rock. Xanthan appears **not** to be strongly adsorbed onto many reservoir rocks. It may even be excluded from some of the rock surfaces owing to repulsive forces. Close to the walls of the rock pores, there may consequently be a layer of reduced polymer concentration, thus increasing the flow of polysaccharide molecules through the interstices in the rock. By contrast, polyacrylamide tends to adsorb to the rock surface and the effective concentration in solution is rapidly reduced.

The use of biopolymers in oil production exploits the ability of the poly-saccharides to modify the properties of the aqueous solution. Some of the properties utilised are the same as those employed in other industrial applications such as water-based thixotropic paints, paper and other surface coatings, pharmaceutical and food extrusion, etc. However, the subsurface environment is inimical to the chemicals used whether they are of biological or chemical origin. The reservoir water may contain high levels of total dissolved salts and may range from acid to alkaline pH (Table 10.2). The temperature, especially in some of the deeper offshore reservoirs

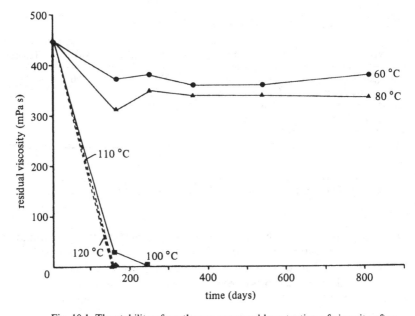

Fig. 10.1. The stability of xanthan as measured by retention of viscosity after storage at high temperature in the absence of oxygen. Residual viscosity was measured at a temperature of 30 °C and a shear rate of 1 s^{-1}. (Reproduced from Kierulf and Sutherland, 1988).

of the UK and Norwegian Continental Shelf, may be high, typically in excess of 80 °C. Despite this, testing of biopolymers such as xanthan, scleroglucan and succinoglycan has shown that, *under suitably controlled conditions*, both xanthan and scleroglucan can retain their structure and physical properties in the temperature range of 80–90 °C, for periods in excess of 800 days (Fig. 10.1). Although exopolysaccharides have found use in the shallower, relatively cool and much smaller reservoirs found onshore in North America, and have been tested onshore in some small oilfields in North Germany, their application in hotter, deeper reservoirs is still being evaluated. Such evaluation, especially for some of the smaller and better defined offshore reservoirs with temperatures near 80 °C, indicates a polymer requirement of 8.7 lb (3.95 kg) per barrel of incremental oil. The properties of exopolysaccharides that have to be determined in assessing their suitability for EOR usage are indicated in Table 10.1. A problem associated with their use in the oil industry, however, is that formulation on site may be poorly controlled. The injection water must be prepared with continuous mixing and quality control is difficult.

Profile modification

Microbial polysaccharides have found a significant market share in drilling muds and completion fluids. There are also convincing arguments for their use as mobility control agents in water injection for enhanced oil recovery. A further development is in the area of selective pore and structure blockage to modify the reservoir profile. Oil reservoirs are inhomogeneous, usually comprising layers with very marked differences in permeability from zone to zone. High-permeability zones are rapidly swept during water flooding, leaving low-permeability zones unswept. Blocking agents can be used to divert the water flood into these unswept sections of the oil reservoir. A further development, which has recently been proposed, is to use polymer injection to suppress influx from aquifers present in oil-bearing rock formations (Hughes *et al.*, 1988). The use of xanthan cross-linked with chromium ions provides a reliable method of forming a gel structure *in situ*. The rate of gel formation depends primarily on the concentration of Cr^{3+}; the xanthan concentration is less critical. For 50 mM Cr^{3+}, gelation occurs in less than one hour, but at concentrations of Cr^{3+} of 2 mM, about 40 h are needed (Lund *et al.*, 1988). The creation of the xanthan gel effectively blocks high-porosity and similar 'thief zones', which would otherwise lead to considerable loss of mobility control fluids and rapid breakthrough of water into the produced oil. The cross-linked gels are relatively insensitive to salts and have sufficient strength to withstand water flow. They can also be readily broken and removed should it prove necessary to regain access to the high-permeability zone.

The system is premixed on the surface and, because of its pseudoplasticity, can be pumped into place. When injection is completed, the reduced shear forces permit restoration of viscosity and the development of the gel. If access to the high-permeability zone is later required, the gel can be destroyed by treatment with an oxidising agent such as sodium hypochlorite. Field-scale trials have resulted in claims of substantially altered flow patterns and significant incremental oil recovery.

Because of the need to penetrate reservoir rock structures with low permeability, biopolymers for use in enhanced oil recovery have to be free from particulate matter and thus of much higher quality than those provided for drilling purposes. The particles would otherwise block rock structures and reduce recovery. Particulate material may either be cell-derived (e.g. residual microbial cells) or may be in the form of 'microgels', polysaccharide aggregates present in exopolysaccharide preparations which have been dried and redissolved in aqueous solution. The microgels may be in the form of polysaccharide molecules that have been effectively cross-linked to form gels by multivalent cations present in the preparation. Alternatively, larger aggregates of unhydrated or partly hydrated polymer may be present if the shearing applied during rehydration of dry xanthan powder has been inadequate. Another possibility is the presence of multimolecular aggregates, which are readily deformable and thus can penetrate to some extent into rock formations when under high pressure gradients. Once the pressure gradient is reduced, they regain their previous form and cause plugging inside the formation. It has recently been

Table 10.3. *Annual oil-well usage of polymers (tonnes)*

Products	Europe	N. America	Middle/Far East
cellulose derivatives	12500	44000	16400
guar and derivatives	2250	5000	2100
xanthan	1500	10000	2000
polyacrylamide	3000	10000	2000

suggested that microgel formation results from association between the polysaccharide and denatured protein in the presence of salts. These intramolecular interactions provide foci for the development of microgels.

Sequential treatment with endo-β-glucanases and alkaline proteases serves to clarify the polysaccharide solutions and greatly increase their injectivity; this could well be attributable in part to the destruction of protein foci. Enzyme treatment can be applied either at the site of microbial exopolysaccharide production or at the point of application, without greatly increasing the product and process cost. Certain of the newer xanthan products are also relatively free of particulate material through the use of improved purification and membrane recovery processes.

The biopolymers that have so far shown the greatest potential for enhanced oil recovery applications are xanthan and scleroglucan. Both have been shown to be relatively stable at high temperature and, unlike the polyacrylamides, are not affected by high salt concentrations. Rather, the presence of relatively high concentrations of cations promotes the ordered polysaccharide conformation, raises the transition temperature and enhances polymer stability. However, oxygen must be rigorously excluded from the solution prior to injection as it tends, particularly in the presence or iron, to promote free-radical reactions. These lead to extensive and rapid degradation of the biopolymers. In laboratory experiments, xanthan samples have shown good retention of viscosity for periods of up to 800 days at 90 °C. Higher temperatures result in rapid degradation of the polysaccharide (Fig. 10.1).

New exopolysaccharides with higher thermal stability than xanthan may well be found. Indeed there is some evidence that certain of the polymers belonging to the 'gellan' family (pp. 27–8) are probably more thermostable than xanthan. However, they are also of low injectivity, which precludes their use in enhanced oil recovery operations.

The recent volatility of the oil industry, and the low prices obtained even for high-grade crudes, have had a very marked effect on drilling activity. This has been much more noticeable in terms of exploration activity than in the drilling of production wells. Following the drilling of appraisal wells, a number of new fields are being brought on stream. The depressed oil price has had an even greater effect on the introduction of enhanced recovery techniques. There is at present little financial incentive for the use of biopolymers and similar relatively high-cost chemicals for enhanced oil recovery. Adoption of tax incentives for enhanced oil production, a fiscal policy adopted in the USA, may eventually alter this situation in other parts of the world. The estimated market share of xanthan for oil usage in the three major oil-producing areas of the world can be seen in Table 10.3.

10.2. Polysaccharide use in enzyme technology

Enzyme extraction in two-phase systems

Aqueous two-phase systems provide an elegant mechanism for the recovery and purification of biological molecules, including enzymes. In biotechnology, the two-phase systems have found application in the large-scale purification of enzymes and interferon, as well as in affinity purification. Aqueous phase systems for the cultivation of enzyme-producing microorganisms have also been described. The technique requires the formation of two distinct phases from an aqueous solution and the high solubility of enzymes in the polymer solutions. They may either contain two water-soluble polymers or one such polymer and a high concentration of a phosphate or sulphate salt. Among the systems that have been extensively studied are solutions containing the synthetic polymer polyethylene glycol (PEG) in combination with either dextran or potassium sulphate. Using such techniques, extensive purification can be achieved.

Dextran has, however, some disadvantages. Crude dextran is inexpensive but its high molecular mass results in highly viscous solutions in the lower phase of dextran–PEG mixtures. Alternatively, purer, low-molecular-mass dextran yields lower viscosity but is considerably more expensive. An alternative exopolysaccharide for use in two-phase systems is *pullulan*, which can be commercially produced at a lower price (US$11.5 per kilogram). The phase diagrams obtained from PEG–pullulan mixtures and the phase partition of several enzymes depended on the molecular mass of the PEG used. Very similar diagrams were obtained by using PEG 14000 and pullulan, or PEG 8000 and dextran (Fig. 10.2). As well as its use for the purification of enzymes, the system was applied experimentally to separate enzyme products in the upper phase from the microorganisms yielding them in the lower phase (Nguyen *et al.*, 1988).

Enzyme and cell immobilisation

Polysaccharide gels are widely used for the immobilisation of enzymes and cells (Table 10.4). This has the major advantage of physically separating a biocatalyst from the product and of permitting re-use of the catalyst. It may also permit the adoption of a continuous flow system and the use of high cell densities within a reactor. Design of the gel has to ensure that access and

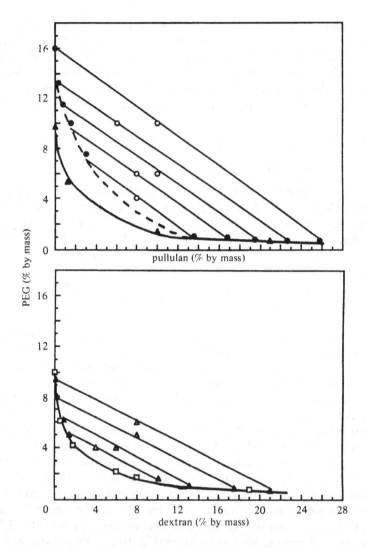

Fig. 10.2. Phase diagrams of polyethylene glycol – polysaccharide mixtures. (Reproduced from Nguyen *el al.* (1988) with permission.)

egress of substrate molecules is satisfactory and that the polymer used has no deleterious effect on cell viability or enzyme activity. For this reason, biopolymers are much preferable to polyacrylamides. The gel may be formed by the addition of ions, as with calcium alginate, or the enzymes may be chemically coupled by a polysaccharide such as carrageenan. These polysaccharides can be used to immobilise cells in a single-step process, under very mild conditions. The cells thus remain viable and enzymes are

Table 10.4. *Immobilised cells*

Cell type	Product
Bacteria	
Anabaena sp.	ammonia
Erwinia sp.	isomaltulose
Lactobacillus bulgaricus	lactic acid
Synechococcus sp.	glutamate
Zymomonas mobilis	ethanol
Fungi	
Aspergillus niger	citric acid
Aureobasidium pullulans	glucoamylase
Claviceps purpurea	alkaloids
Saccharomyces cerevisiae	ethanol, glycerol
Streptomyces spp.	antibiotics
Algae	
Dunalliella tertiolecta	glycerol
Plant cells	alkaloids, digitoxins
Animal cells	monoclonal antibodies insulin

Note:
Further examples are to be found in Skjak-Braek &
Martinsen (1989).

not denatured. The disadvantage of alginate gels, despite the mild
conditions under which they can be formed, is the gradual leakage of
calcium ions, leading to dissolution of the gel. Unless the gel-promoting
ions are carefully added, the gels may be inhomogeneous. Alginate gels
have, however, been very widely employed in immobilisation of microbial
and eukaryotic cells and enzymes. The technique is easy to perform and safe
under laboratory conditions and may even be suitable for maintaining cell
protoplasts without extensive lysis occurring. Cells can be suspended in a
sodium alginate solution (2% final concentration), then extruded into a
solution of calcium chloride to form beads. Algal alginate has been
employed for the studies reported so far, but some bacterial alginates might
also prove suitable. The quality of alginates from either algal or bacterial
sources can also be improved through use of the *polymannuronate
epimerase* enzyme produced by *Azotobacter vinelandii* (p. 67). As alginates
form strong complexes with polycations, compounds such as chitosan,
which do not dissolve in the presence of calcium chelating agents, can be
added to stabilise the gels and to reduce their porosity. Newer products
such as gellan also have considerable potential in this area; XM6, with its

gel–sol transition resembling gelatine, may also be of use for specialised applications at lower temperatures where harsher conditions must be avoided. Gellan has a lower ion requirement for gelation and may thus offer considerable advantages over alginates, especially with regard to long-term stability and the avoidance of leakage of cells or proteins from the immobilised complex. If leakage is substantial, there is increased cost in down-stream processing, thus reversing one of the major advantages of cell or enzyme immobilisation.

The applications of gel-immobilised cells are very diverse (Table 10.4). Immobilised preparations of *Saccharomyces cerevisiae* are capable of producing ethanol from glucose for over 20 days. A particularly intriguing application of alginate-immobilised *S. cerevisiae* cells is their application in the secondary fermentation required for the 'methode champenoise'. Gel beads containing the yeast cells are added to the bottles and ferment the added sucrose to provide ethanol and carbon dioxide. When this process is completed, the beads can be removed much more readily than the cells used traditionally.

Hybridoma cells can be encapsulated within alginate–polylysine beads. This gives a high cell concentration internally, which is separated from proteins in the medium. Different types of cells can also be co-immobilised. Thus *Aspergillus niger* in combination with *Zymomonans mobilis* provide a means of producing ethanol through use of the amylase from the fungi and the ethanol-synthesising enzymes of the bacteria. It is also possible to co-immobilise plants with inoculants. Examples are leguminous plants and *Rhizobium* inoculants, or tree roots and symbiotic mycorrhizal fungi. No single gel system is likely to be ideal for all; different gel-forming polysaccharides will be needed.

As well as using polysaccharides or their derivatives in the purification of enzymes, it has been suggested that they can render enzymes more stable. Proteins can of course be immobilised, but it is also possible to complex them to *soluble* polymers such as dextran. This extends the effective life of the enzymes and makes them more stable to thermal and other forms of denaturation, through effecting a conformational change in the protein.

10.3.   Polysaccharide films

Many polysaccharides can form tough flexible films when carefully dried from aqueous solution. This property may have application in the production of films or coatings that can be used to prolong the shelf-life of fruit and other foodstuffs. Several polysaccharides have been proposed for use in manufacturing films impervious to oxygen for this and similar purposes. These include both pullulan and elsinan (Fig. 3.3). The films prepared from pullulan are strong and resilient and can be readily moulded into different shapes.

10.4.   Polysaccharide use in paints, printing and textiles

The application of an aqueous solution of a dye to fabric or paper results in spreading and loss of definition of any design. The solution has therefore to be thickened. This has two aims: to restrict the flow of the solution, and to retain it within the printed areas until fixation of the dye into the fibres is complete. The polysaccharide thickeners used in the printing industry must confer the appropriate rheological properties on the print paste, but they must also possess other characteristics. After application of the dye solution, a film is formed around the fibres, where the fluid has penetrated the fabric. The film must adhere strongly and uniformly to the fibres during the subsequent process of dye fixation. Finally, any unfixed dye and other chemicals must be removed by washing. Thus, the thickener must swell and dissolve rapidly. Because of their physical properties and compatibility with various dyes and inks, alginates have found wide applicability for textile printing. Xanthan is also used for the printing of both paper and textiles. It is used as a suspending agent and stabiliser for both water-based and emulsion inks and for dye pigments, providing controlled penetration and water release. It has also proved extremely useful for controlling dye rheology in jet printing of carpets and similar materials (Table 10.5).

A number of microbial polysaccharides, including xanthan and succinoglycan, have been employed in the manufacture of paints and pigments. In these products, the ability of the polysaccharide solution to provide stable suspensions of the pigment, combined with the pseudoplasticity which facilitates pumping or spraying, can provide superior properties when compared with other formulations.

10.5.   Polysaccharide gels

One of the non-food uses for mixtures of xanthan and plant galactomannans is in deodorant gels. The mixtures are also used in photographic processing. Mixtures of curdlan and activated carbon provide high absorbing activity without the problems associated with very fine carbon particles. Gels are also used in the preparation of microbiological culture media. The traditional gelling agent for this is agar but is has the disadvantage of problems of supply and provides a gel which is not water clear. An alternative is gellan (gelrite), which gives very clear gels at concentrations rather lower than those of agar. The bacterial polysaccharide disperses and hydrates readily in hot or cold water and forms a gel structure which is suitable for microbial growth. The media prepared with gellan as gelling agent can be autoclaved without any deterioration. Autoclaving can be repeated. Gellan is less easily degraded by enzymes than is agar. It may have potential value for the study of marine microorganisms, many of which are capable of degrading agar.

Table 10.5. *Industrial applications of xanthan*

Usage	Physical properties required
explosives (package gels)	compatibility with $Ca(NO_3)_2$; water resistance (for dynamite)
fire fighting	foam stabilisation
flowable pesticides	suspension and drift control
hydraulic fracturing	viscosity and cross-linking
jet printing	
laundry chemicals	suspension of starch/solids
liquid fertilisers and herbicides	suspension
liquid feed supplements	suspension
oil-drilling muds	shear-thinning and viscosity control
paper finishing	suspension of clay coatings
thixotropic paints	stabiliser
water clarification (ore extraction)	flocculant

10.6.   Horticultural and agricultural applications

Among other possible applications for microbial exopolysaccharides, it has been proposed that they might usefully be added to soil to improve texture. One such product for which this usage is anticipated is the polymer from *Chlamydomonas* species. Whether this can be done economically, given the cost of polysaccharide production, is not clear. Most of the other products from algal production are of much higher value than polysaccharides (for example, β-carotene sells for US$750–1000 per kilogram), but it might be possible to recover the extracellular polymers at an economic cost as by-products during the preparation of other saleable products.

Polysaccharides such as alginate find application as coating agents when young trees or shrubs are planted. The roots are dipped in a solution of polysaccharide to provide polymer film protecting them from desiccation while the seedling is being handled. Coating in this manner may also be of value when employing biological control agents or rhizobial inocula either to plants or to seeds. A more esoteric application involves the dipping of Christmas trees in a polysaccharide solution to prolong the period during which they can be kept in a hot, dry indoor environment before the needles start to fall.

Polysaccharides are also finding increased use in seed coatings and in the manipulation of 'propagules' and young cuttings of plants and shrubs. This includes the supply of packs containing rooting hormones ready prepared and sterilised for plant propagation. A similar suggested use for curdlan involves its inclusion in germination media for young rice plants. A further advantage of using polymers as seed coatings is that legumes can be coated with appropriate inocula of *Rhizobium* species. This material is then used

Table 10.6. *Applications of emulsan*

Viscosity reduction of heavy crude oil (for transport,
   pumping)
Cleaning of tanks and vessels
Concentration of multivalent cations
Dispersion of slurries and of pigments
Enhanced oil recovery from oil sands and shale
Emulsion stabilisation
Flocculation of bentonite and other clays

---

for fluid drilling, in which pregerminated seedlings coated with the gelatinous matrix are planted. Pesticides and nutrients can also be incorporated into the gels. As yet the process has been restricted to small-seeded agriculture and horticultural plants. Among the gels used have been xanthan–galactomannan mixtures and alginates.

10.7. Biosurfactants

Although a number of compounds with emulsifying activity have been isolated from microorganisms, particularly those capable of growth on hydrocarbons, few of these molecules are polysaccharide in nature. An exception is *emulsan*, produced by *Acinetobacter* RAG1. This complex polymer resembles classical lipopolysaccharides in some aspects of its structure (p. 27), has an average relative molecular mass of 9.9×10^5 and an intrinsic viscosity of 505 $cm^3 g^{-1}$. Emulsan enhances the formation and stability of a wide range of hydrocarbon-in-water emulsions, including mixtures of water with kerosine, gas oil or crude oil. Native emulsan containing *ca.* 20% protein is more effective than *apoemulsan* (less than 5% protein content) in emulsion formation. The emulsan molecules bind tightly to the surface of hydrocarbon droplets and the stabilised droplets can be separated from the bulk water phase by centrifugation.

Despite its structural resemblance to lipopolysaccharides, emulsan has very low toxicity and it has been proposed for use in combination with other surfactants as an emulsifying agent. Very low concentrations of emulsan in the range 0.1–1.0% were required. A range of industrial applications has been proposed (Table 10.6). It has also been incorporated into more specialised formulations such as hand-cleaning materials developed specifically for the removal of compounds used in paint, plastic and printing ink manufacture. These emulsan hand-cleaners are thought to be particularly valuable in removing paints, lacquers and printing inks and, if used sufficiently quickly, are even effective in removing polyurethane paints incorporating catalytic hardeners. Other exopolysaccharides without the highly lipophilic attributes of emulsan are nevertheless useful in the preparation of stable emulsions, especially in the cosmetic and pharmaceutical industries.

Further reading

Albertsson, P. A. (1986). *Partition of cell particles and macromolecules* (3rd edn). Wiley Interscience, New York.
Johansen, R. T. & Berg. R. L. (1979). *Chemistry of Oil Recovery*. ACS Symposium **91**. American Chemical Society, Washington.
Tampion, J. & Tampion, M. D. (1987). *Immobilized cells: principles and applications*. Cambridge University Press.

11 Medical applications of exopolysaccharides

Exopolysaccharides have a number of industrial applications which relate either directly or indirectly to medicine. One such is the use of dextran and its derivatives on both a laboratory and an industrial scale for the purification of compounds of medical interest, including pharmaceuticals, and of enzymes for diagnostic purposes. Polysaccharides may also be used to encapsulate drugs for their gradual delivery and they may be used to immobilise enzymes employed for diagnosis or for the chemical modification of pharamaceutical products. These applications clearly utilise the *functional* properties of the polysaccharides, such as their rheology or capacity for gel formation. Alternatively, the pharmacological or other *biological* properties of the polymers may be employed. There are essentially three types of direct application of the biological properties of exopolysaccharides to medicine. The exopolysaccharides may be used as vaccines in preference to whole microbial cells or cultures. Thus side-effects due to other cell components such as lipopolysaccharides or proteins are avoided. On the other hand, not all polysaccharides are good immunogens, nor are the exopolysaccharides necessarily the major factors in the specific disease syndromes caused by the polysaccharide-producing microbial pathogens.

Several exopolysaccharides mimic eukaryotic polymers in their structural details. For this reason, they may be associated with certain specific diseases such as meningitis (in the case of sialic acid-synthesising bacteria). Some of these polysaccharides are, however, useful as substrates for determining enzyme specificity, and others are used as substitutes for the eukaryotic polymers. Finally, a number of exopolysaccharides having antitumour or antiviral activity have been identified.

11.1. Pharmaceutical applications

Polysaccharides have a number of applications in the formulation of pharmaceutical products. They are incorporated into lotions and gels to impart specific rheological properties to the preparations. The materials used are mainly of plant or algal origin, including alginate, but various microbial polysaccharides, including xanthan, may also find such uses. The properties utilised are essentially the same as those employed in food preparations and also take advantage of the 'mouth-feel' and lack of flavour of the polymers. The shear-thinning capabilities of xanthan have caused it to be used in some toothpastes, where it permits the product to be readily squeezed from a tube and regain its viscosity on leaving the container. In this role, the flavour release by the bacterial polysaccharide is also important. New gel-forming microbial products such as gellan clearly have potential uses in the pharmaceutical area, as do polymers capable of forming coacervates with gelatine, etc.

11.2.   Wound management

Dextrans are used as plasma substitutes. When used thus as blood expanders, the polysaccharide fractions must be within the relative molecular mass range 75 000 ± 25 000. Products of lower molecular mass are eliminated too rapidly from circulation to be of therapeutic value, and larger polymers interfere with the coagulation of blood. The dextrans are used at a concentration of *ca.* 6%, which yields viscosity and colloid-osmotic behaviour similar to human blood plasma. They are not readily degraded and are less likely to cause liver damage than are other plasma substitutes. As dextrans are resorbed well into body tissue, they can be used for treatments involving the supply of iron during anaemia.

One novel application of algal alginates may also be applicable to microbial polymers. *Calcium alginate* fibre can be made by wet spinning, using a solution of sodium alginate passed into a bath of calcium chloride solution. The calcium alginate filaments are washed to remove excess calcium; they are of moderate strength but are susceptible to moisture. They can be manufactured into wound dressings, which absorb fluid from wound exudates, are non-irritant and haemostatic. The dressing can finally be removed easily and painlessly without damaging the scar tissues and causing further trauma. Microbial alginate with appropriate composition could be used in place of algal products, but there is also potential for developments of mixed and other new products with novel properties. This is seen in the first commercial use of bacterial cellulose in novel wound-dressing products, forming an artificial 'skin' over serious burns.

Hyaluronic acid is a natural constituent of connective tissue and can be used as a joint lubricant. It is also available for use in eye surgery, where it finds application because of its apparent identity to the material found in the vitreous humour. A further application, relating to this, is its inclusion in artificial tears prescribed for patients with defective tear ducts. It is also an ingredient of various pharmaceutical preparations, including lotions for skin 'rejuvenation'. Hyaluronic acid may eventually also find applications in preparations for drug delivery and for wound healing.

11.3.   Polysaccharide vaccines

The very diversity of polysaccharides may diminish the possibility of using them as vaccines. There are frequently too many different chemotypes for it to be practicable to prepare a potential vaccine. However, several bacterial diseases are caused by a small number of polysaccharide-producing serotypes as the dominant causes of infection. It is then possible to evaluate such polymers to determine whether they might be suitable for vaccine use. The antibody response to polysaccharide antigens varies greatly and can be affected by a number of factors including both chemical and physical properties of the exopolysaccharides. The specific chemical structure of the polysaccharide clearly plays a major role. In addition, the conformation in solution and the molecular mass are likely to affect antigenicity. As the immune response to polysaccharides is T-cell independent, this accounts for the poor immunogenicity of most polysaccharides. There is a lack of memory response and in humans a late response occurs. Thus the population most at risk to several infections caused by capsulate bacteria, infants of less than two years, shows a very poor immune response. Another potential problem is the possible role of genetic factors in regulating immune response to bacterial polysaccharides. This may affect the susceptibility of certain infants to infections caused by polysaccharide-producing bacteria (Lenoir *et al.*, 1988).

A problem associated with exopolysaccharide preparations from Gram-negative bacteria is the possible presence of small amounts of lipopolysaccharide derived from the bacterial walls. This can cause pyrogenicity to an extent sufficient to preclude the use of the exopolysaccharides as components of multivalent vaccines for human use. Such pyrogenicity resulting from lipopolysaccharides can be reduced by deacylation in the presence of mild alkali. This treatment may cause some degradation of the exopolysaccharide and perhaps some reduction of its immunogenicity. In particular, ester-linked acyl groups will be lost. Better vaccines can be obtained when the polysaccharides are coupled to protein. This converts the response to T-cell dependency and makes use of the ability of young infants to respond to protein antigens owing to their recognition by the T cells. Several systems have now been examined.

Klebsiella species can constitute a major threat to patients who have been immune-compromised, especially as multiple antibiotic resistance may be prevalent among such bacteria. Laboratory tests with a number of different serotypes evoke an anticapsular immunoglobulin G response, although not all the polysaccharides were equally immunogenic. Human immunoglobulin G against serotype 1 exopolysaccharide, prepared from volunteers, was also effective in preventing fatal sepsis in laboratory animals. A safe immunogenic vaccine may thus be developed from those *Klebsiella* species that present clinical problems in certain groups of patients (Allen *et al.*, 1987*a*, *b*).

Some of the microbial exopolysaccharides that might be proposed for vaccine development are very poor immunogens in humans. This is the case with the group B meningococcal polysaccharides and the structurally identical products from *Escherichia coli* K1. Both are 2,8-α-linked sialic acid homopolysaccharides (p. 24). In colominic acid from the *E. coli* strain, the poor immunogenicity could in part be related to the low average chain length of 16–17 monosaccharides. However, the group B meningococcal polymer is larger, with a chain length of 190–200. Clearly chain length in itself is not the determining factor in the immunogenicity of these exopolysaccharides, although conformation may play a role. By contrast, *N. meningitidis* group C sialic acid has an average chain length of about 320 and, like the group B material, possesses phosphatidyl diglyceride moieties that are covalently bound to the reducing termini of the polysaccharide chains. The poor immunogenicity is more probably due to immunological tolerance of tissue components with structurally identical sialic acid-containing glycolipids. The structural similarities between the bacterial product and the host material probably also account for the role of the former as virulence factors in bacterial meningitis. A possible solution to this problem has been explored through the chemical modification of the *bacterial* sialic acids. Coupling of the polysaccharides to protein carriers failed to provide a suitable immunogen. As an alternative procedure, chemical modification has been attempted, the aim being to obtain a simple procedure and avoid chemical degradation of the bacterial polysaccharides. The method finally adopted involved the removal of *N*-acetyl groups and their replacement with *N*-propionyl groups. As this also involves a reduction in the molecular mass of the bacterial polymers, the modified sialic acids were coupled to tetanus toxoid. In animal systems this approach yields a good artificial immunogen, which induces antibodies cross-reacting with group B meningococcal polysaccharides and killing the bacteria in the presence of complement.

In the USA, three bacterial polysaccharide vaccines are currently available. These are: (i) *Haemophilus influenzae* type b; (ii) meningococci belonging to serogroups A, C and Y; and (iii) a multivalent vaccine for *Streptococcus pneumoniae* (pneumococcus). These bacteria are primarily causative organisms of acute diseases of infancy and a problem relating to the use of such polysaccharide vaccines is the poor response elicited in young children under the age of 2 years, the group who are often most at risk from the infections caused by these bacteria. Attempts are therefore frequently made to improve the antigenicity of the polysaccharides through coupling to proteins or other chemical modification. Control of these serogroups of meningococci is however now possible, at least in the short term, through the use of purified exopolysaccharide material in vaccines. The *H. influenzae* type B strains are also causative organisms of meningitis. These highly virulent bacteria form one of six antigenically and chemically distinct capsular chemotypes. Although antipolysaccharide antibodies provide protection against infection, the polymer is poorly antigenic and does not induce protective immunity in infants of two years old or less (the most susceptible group to *H. influenzae* meningitis). Immunogenicity can be enhanced through coupling to proteins; it has been suggested that a protein from the outer membrane of the causative microorganism should be used. The capsular polysaccharide is a linear heteropolymer composed of alternating units of ribose and ribitol linked by a phosphodiester group, which has proved amenable to coupling to protein.

11.4. Antitumour and antiviral activity

Non-specific protective activity against a number of viruses has been demonstrated in the laboratory for several polysaccharides. These were mainly polymers from eukaryotes and included heparin and carrageenan, potent inhibitors of herpes and other animal viruses. The polysaccharides were without cytotoxic effects and their mode of action appeared to be that of blocking viral attachment. They were only effective during the early stages of viral attachment and development. The antiviral activity was not obviously dependent on structure of the polysaccharide, as the effective polymers ranged from heparin to carrageenan, dextran sulphate and alginic acid, a structurally diverse group of neutral and polyanionic products. Microbial exopolysaccharides may well show similar effects, particularly those which are structurally related to some of these eukaryotic polymers.

A feature of a number of 1,3-β-D-glucans of fungal origin is their antitumour activity. The polysaccharides which demonstrate this activity are all glucans which are closely related in their structure to scleroglucan (Fig. 3.1), but vary in their water solubility and in the degree and nature of their side-chains. Because of their relatively low water-solubility, some of these polymers have to be isolated from the fungal fruiting bodies or culture filtrates with hot water or alkali. Antitumour activity depends on the source of the polysaccharide. Material from *Ganoderma lucidum* showed high activity at a dosage of 10 mg kg^{-1} against Sarcoma 180 tumour in mice. Activity was higher than that of other fungal β-glucans, which possessed a very high degree of branching, but less than that obtained when using a polymer from *Volvariella* (Table 11.1). Polysaccharides modified by periodate oxidation, followed by borohydride reduction to yield glucan polyols, showed even higher antitumour activity. Further modification of the derived material, a fraction of relative molecular mass 4.7×10^5 produced by prolonged ultrasonic treatment, increased the antitumour activity still further. This material was effective against both allogenic and syngenic mouse-implanted tumours at doses of 1–5 mg kg^{-1} by intraperitoneal administration.

The bacterial polysaccharide curdlan, which lacks the glucosyl side-chains of scleroglucan, also demonstrates antitumour activity *in vivo* at doses similar to those used for the fungal β-D-glucans. In common with the fungal β-glucans, it can form triple-helical structures in the ordered

Table 11.1. *Microbial sources of β-glucans with antitumour activity*

Species	Degree of branching
Volvariella	1 : 4
Ganoderma lucidum	1 : 2.7
Pestalotia spp.	1 : 1.6
Auricularia auricula-judae	1 : 1.2
Grifolia umbellata	not known accurately
Schizophyllum commune	not known accurately
Agrobacterium sp[a]	0
Alcaligenes faecalis var	
myxogenes[a]	0

Note:
[a] Curdlan.

conformation. Curdlans of different molecular mass have been used, as well as carboxymethyl- and other curdlan derivatives.

It has been suggested that sulphated polysaccharides of non-microbial origin inhibit tumour metastasis by blocking tumour-developed heparanases, but this mode of action has not been confirmed for the triple-helix-forming β-glucans. As a source of sulphated *bacterial* polysaccharide has now been found in cyanobacterial species, it will be of interest to determine whether such polysaccharides possess antitumour activity comparable to that of the sulphated eukaryotic polymers. Apart from the adoption of triple-helical conformation in aqueous solution, little is known about the physical requirements needed to determine antitumour activity. Some of the fungal polysaccharides, both in the native form and after conversion to the polyol form, are of very high relative molecular mass (over 2×10^6). Curdlan is also normally of very high molecular mass. However, schizophyllan varies in relative molecular mass from 5×10^3 to 1.3×10^5. It has been noted that material of relative molecular mass less than 10^4 is ineffective in experimental tumour treatment. This correlates with the decreased capacity of this fraction to form triple helices and with a high proportion of single chains in aqueous solution. Similar results were obtained using curdlan of lower molecular mass. At present, it is not clear whether this group of microbial exopolysaccharides has any value in clinical treatment of tumours or whether their value is limited to experimental studies.

Protein-bound polysaccharides from basidiomycetes have been tested clinically as immunomodulators. They produce antitumour activity against some tumours, probably through the restoration of the depressed delayed-type hypersensitivity or through antibody production. A further physiolo-

gically active polysaccharide from *Bacillus polymyxa* reduced the choles-
terol levels of hypercholesterolamic laboratory animals. The polymer responsible for this activity was thought to contain neutral sugars (D-glucose,
D-galactose and D-mannose) together with both D-glucuronic acid and
D-mannuronic acid, but no structural details or physical characteristics
that might explain this unusual activity have been reported. The same
property has been demonstrated with various polysaccharides of plant
origin, such as pectin and guar gum. The proposed mechanism for this
phenomenon has been that of lowering the assimilation of bile salts; not all
polysaccharides exhibit this effect.

Polysaccharides administered intravenously or intraperitoneally have
also been shown to protect against some of the effects of ionising radiation.
Experiments in mice indicate that the recovery of bone marrow cells is
enhanced and that nucleated cells in the spleen are preserved. Significant
protection has been found by using polysaccharides extracted from
Saccharomyces cerevisiae and other yeasts, possibly through protection of
bone marrow stem cells. The polysaccharides tested are homopolysaccharides: mixed-linkage glucans or mannans. It is not yet clear whether
exopolysaccharides, especially those with antitumour activity, also possess
the same or similar effects.

11.5.   Pharmacological applications

Heparin, obtained from eukaryotic material, is widely used as an
antithrombotic and antilipaemic drug. Attempts have been made to
improve the therapeutic efficiency of heparin and to reduce side-effects,
through modification of the chemical structure and the molecular size. An
alternative might be to obtain exopolysaccharides of microbial origin, such
as the product of *Escherichia coli* K5, which have close structural similarity
to heparin. It is also possible that oligosaccharides derived from microbial
sources, including the cyclic β-glucans from *Rhizobium* species, may prove
useful as the basis for synthesis of new material with pharmacological
activity. These polysaccharides and oligosaccharides, including those with
chiral function, may also find applications as affinity adsorbents in the
purification of pharmaceutical products. Alginates and perhaps other
polysaccharides such as gellan, have possible uses in the development of
encapsulated products for the slow release of drugs or hormones. It is thus
probable that an increasing number of novel medical applications will be
found for exopolysaccharides in the coming years, as for example, the use of
dextran-ferrite magnetic particles for use in NMR imaging.

Further reading

Bell, R. & Torrigiani, G. (1987) *Towards better carbohydrate vaccines.*
Wiley, Chichester.

Crescenzi, V., Dea, I. C. M., Paoletti, S., Stivala, S. & Sutherland, I. W. (1989). *Recent developments in industrial polysaccharides: biomedical and biotechnological advances.* Gordon and Breach, New York.

Timmis, K. N., Boulnois, G. J., Bitter-Suermann, D. & Cabello, F. C. (1985). Surface components of *Escherichia coli* that mediate resistance to the bactericidal activities of serum and phagocytes. *Current Topics in Microbiology and Immunology* **118**, 197–218.

12 The future for microbial exopolysaccharides

If one looks at the various areas in which microbial exopolysaccharides are currently employed, it may be possible to make some predictions about future usage. The increase in interest in the physical properties of these polymers, together with a much better understanding of the relationship between physical properties and chemical structure and the continued search for new polysaccharides, will inevitably lead to new discoveries. Relatively few of these are likely to have properties suited to new applications or their use in place of currently used polymers. There are two major constraints: legislative and financial.

In the food industry, xanthan is *currently* unique in its acceptability. As has been mentioned earlier, gellan from *Pseudomonas elodea* is currently undergoing safety evaluation. These two polymers can potentially fulfil many of the perceived needs of the food industry for microbial polysaccharides as well as replacing some established plant or algal products. Any new polymer would only have a small market niche and this would probably be insufficient to justify the expense of development and of the safety appraisal needed to obtain legislative approval. It is more likely that **new applications** will be found for the polysaccharides, such as xanthan, which already are approved. An exceptional situation exists in Japan, where the microbial polysaccharides, being regarded as *natural* products, are acceptable food ingredients. *Perhaps* some new polysaccharides *will* find applications in Japan, which may justify their introduction into other countries. One such might be curdlan.

In non-food applications, xanthan currently holds a commanding situation. Although it is in chemical terms an expensive product, it will probably remain unchallenged by other microbial polymers in terms of cost. Economies in the production process may even be able to **reduce** the cost slightly, although the margin for cost reduction is probably small. If it finds widespread use in enhanced oil recovery, there *could* be greatly increased production in a number of different geographical locations. This probably awaits realistic financial incentives for incremental oil recovered, as well as improved reservoir modelling and predictive techniques. Other polymers with even greater thermal resistance and *in situ* stability may well complement xanthan for EOR and other oil industry applications. Indeed, the **technological usage** of microbial exopolysaccharides is a definite area for potential growth. This is partly due to the numerous studies on physical properties, but also results from new techniques.

Table 12.1. *Uses and costs of exopolysaccharides*

Polymer	Estimated annual consumption (tonnes)		Cost (US$/kg)
	USA	worldwide	
alginate		23000	5–15
curdlan			940
dextran	2000		35–390
dextran derivatives	600		400–2800
gellan	—[a]		66–75
hyaluronic acid	500		2000–100000
pullulan			11.5
rhamsan gum	—[a]		25
welan gum	—[a]		25
xanthan	20000		10–14
gum arabic		25000–40000	2.8
gum guar		10000–15000	0.9
gum tragacanth			25

Note:
[a] Not yet commercialised, although now available on a trial basis.

Introduction of new polysaccharides may thus occur in the industrial area. Whether there remain large potential areas of application is not clear. Nor is it clear whether any new polymers will result from a deliberate search for polysaccharides with the necessary properties or from the modification of existing products. It is perhaps less likely that new products will come from fortuitously found new microbial isolates. Microbiologists and polysaccharide chemists will need to identify the requirements for new products very clearly; the current optimistic comments so frequently found in scientific papers describing new polysaccharide-producing isolates are seldom likely to be translated into valuable new products.

In the medical area, new polysaccharide or polysaccharide-conjugate vaccines may be developed, but the **amount** of polymer required will remain small. It has, however, to be remembered that such polymers are more expensive to produce and that the cost of vaccine evaluation is high. A more probable growth area is that of pharmaceutical and pharmacological applications. Again, the amounts needed may well be small although, as can be seen from Table 12.1, the cost of a high-quality product such as hyaluronic acid can be very high indeed. Much may depend on genetic engineering to permit production of these high value products in microbial systems more amenable to standard fermentation procedures, although yields of 6.7 g l^{-1} have been reported from mutant strains of *Streptococcus* *equi* or *S. zooepidemicus* lacking hyaluronidase. New developments utilising the very high water-retention capacity of hyaluronic acid (6 1 g^{-1}) can also be expected, particularly in the cosmetics area.

It should not be forgotten that microbial polysaccharides are biodegradable, they are prepared from renewable resources and they are inherently safe products. These properties alone may lead to their wider use in place of non-renewable materials or in a search for a safer and cleaner environment.
renewable materals or in a search for a safer and cleaner environment.

References

Chapter 1

Anderson, A. N., Parolis, H. & Parolis, L. A. S. (1987). Structural investigations of the capsular polysaccharide from *Escherichia coli* 09:K37:A84a. *Carbohydrate Research* **163**, 81–90.

Bayer, M. E., Carlemalm, E. & Kellenberger, E. (1985). Capsule of *Escherichia coli* K29: Ultrastructural preservation and immunoelectron microscopy. *Journal of Bacteriology* **162**, 985–91.

Osman, S. F. & Fett, W. F. (1989). Structure of an acidic exopolysaccharide of *Pseudomonas marginalis* HT041B. *Journal of Bacteriology* **171**, 1760–2.

Stokke, B. T., Elgsaeter, A. & Smidsrod, O. (1986). Electron microscopic study of single and double-stranded xanthan. *International Journal of Biological Macromolecules* **8**, 217–25.

Zevenhuizen, L. P. T. M. (1984). Gel-forming capsular polysaccharide of fast-growing rhizobia: occurrence and rheological properties. *Applied Microbiology and Biotechnology* **20**, 393–9.

Chapter 2

Lindberg, B. (1981). Structural studies of polysaccharides. *Chemical Society Reviews* **10**, 409–34.

Chapter 3

Philip-Hollingsworth, S., Hollingsworth, R. I. & Dazzo, F. B. (1989). Host-range related structural features of the acidic extracellular polysaccharides of *Rhizobium trifolii* and *Rhizobium leguminosarum*. *Journal of Biological Chemistry* **264**, 1461–6.

Tayama, K., Minikami, H., Entani, E., Fujiyama, S. & Masai, H. (1985). Structure of an acidic polysaccharide from *Acetobacter* sp. NBI 1022. *Agricultural and Biological Chemistry* **49**, 959–66.

Zevenhuizen, L. P. T. M. & van Neervan, A. (1983). Surface carbohydrates of *Rhizobium*. *IV*. Gel-forming capsular polysaccharide of *Rhizobium leguminosarum* and *Rhizobium trifolii*. *Carbohydrate Research* **118**, 127–34.

Chapter 4

Sutherland, I. W. (1987). Xanthan lyases – novel enzymes found in various bacterial species. *Journal of General Microbiology* **133**, 3129–34.

Chapter 5

Freeze, H. H. & Wolgast, D. (1986). Biosynthesis of methylmannosyl residues in the oligosaccharides of *Dictyostelium discoideum* glycoproteins. *Journal of Biological Chemistry* **261**, 135–41.

Frosch, M., Weisgerber, C. & Meyer, T. F. (1989). Molecular characterization and expression in *Escherichia coli* of the gene complex encoding the polysaccharide capsule of *Neisseria meningitidis* group B. *Proceedings of the National Academy of Sciences, U.S.A.* **86**, 1669–73.

Ielpi, L., Couso, R. & Dankert, M. (1981). Pyruvic acid residues are transferred from phosphoenolpyruvate to the pentasaccharide–P–P–lipid. *Biochemical and Biophysical Research Communications* **102**, 1400–8.

Ielpi, L., Couso, R. & Dankert, M. (1983). Xanthan gum biosynthesis: acetylation occurs at the prenyl-phosphate sugar stage. *Biochemistry International* **6**, 323–33.

Jarman, T. R. & Pace, G. W. (1984). Energy requirements for microbial exopolysaccharide synthesis. *Archives of Microbiology* **137**, 231–5.

Kamisango, K., Dell, A. & Ballou, C. E. (1987). Biosynthesis of the mycobacterial *O*-methylglucose lipopolysaccharide. *Journal of Biological Chemistry* **262**, 4580–6.

Lederkrener, G. Z. & Parodi, A. J. (1984). 3-*O*-Methylation of mannose residues. *Journal of Biological Chemistry* **259**, 12514–18.

Chapter 6

Cornish, A., Greenwood, J. A. & Jones, C. W. (1988*a*). Binding-protein-dependent glucose transport by *Agrobacterium radiobacter* grown in glucose-limited continuous culture. *Journal of General Microbiology* **134**, 3099–110.

Cornish, A., Greenwood, J. A. & Jones, C. W. (1988*b*). The relationship between glucose transport and the production of succinoglucan exopolysaccharide by *Agrobacterium radiobacter*. *Journal of General Microbiology* **134**, 3111–22.

Jarman, T. R. (1979). In *Microbial polysaccharides and polysaccharases* (ed. R. C. W. Berkeley, G. W. Gooday & D. C. Ellwood), pp.35–50. Academic Press, London.

Jarman, T. R., Deavin, L., Slocombe, S. & Rhighelato, R. C. (1978). Investigation of the effect of environmental conditions on the rate of exopolysaccharide synthesis in *Azotobacter vinelandii*. *Journal of General Microbiology* **107**, 59–64.

Tait, M. I. (1984). Exopolysaccharide production by *Xanthomonas campestris*. Ph.D. thesis, University of Edinburgh.

Chapter 7

Allen, P., Hart, C. A. & Saunders, J. R. (1987). Isolation from *Klebsiella* and characterization of two *rcs* genes that activate colanic acid capsular biosynthesis in *Escherichia coli*. *Journal of General Microbiology* **133**, 331–40.

Dolph, P. J., Majerczak, D. R. & Coplin, D. L. (1988). Characterization of gene cluster for exopolysaccharide biosynthesis and virulence in *Erwinia stewartii*. *Journal of Bacteriology* **170**, 865–71.

Echarti, C. E. B., Hirschel, B., Boulnois, G. J., Varley, J. M., Waldvogel, F. & Timmis, K. N. (1983). Cloning and analysis of the K1 capsule biosynthesis genes of *Escherichia coli*: lack of homology with *Neisseria meningitidis* group B DNA sequences. *Infection and Immunity* **41**, 54–60.

Kroll, J. S., Hopkins, I. & Moxon, E. R. (1988). Capsule loss in *Haemophilus influenzae* type b occurs by recombination mediated disruption of a gene essential for polysaccharide export. *Cell* **53**, 347–56.

Roberts, I. S., Mountford, R., Hodge, R., Jann, K. & Boulnois, G. J. (1988). Common organisation of gene clusters for production of different capsular polysaccharides (K antigens) in *Escherichia coli*. *Journal of Bacteriology* **170**, 1305–10.

Silver, R. P., Aaronson, W. & Vann, W. F. (1987). Translocation of capsular polysaccharides in pathogenic strains of *Escherichia coli* requires a 60 kilodalton periplasmic protein. *Journal of Bacteriology* **169**, 5489–95.

Torres-Cabassa, A. S. & Gottesman, S. (1987). Capsule synthesis in *Escherichia coli* K12 is regulated by proteolysis. *Journal of Bacteriology* **169**, 981–9.

Vanderslice, R. W., Doherty, D. H., Capage, M. A., Betlach, M. R., Hassler, R. A., Henderson, N. M., Ryan-Graniero, J. & Tecklenburg, M. (1989). Genetic engineering of polysaccharide structure in *Xanthomonas campestris*. In *Biomedical and biotechnological advances in industrial polysaccharides* (ed. V. Crescenzi, I. C. M. Dea, S. Paoletti, S. Stivala & I. W. Sutherland), pp.145–56. Gordon and Breach, New York.

Wang, S-K., Sa-Correia, I., Darzin, A. & Charabarty, A. M. (1987). Characterization of the *Pseudomonas aeruginosa alginate (alg)* gene region II. *Journal of General Microbiology* **133**, 2303–14.

Chapter 8

Atkins, E. D. T., Attwool, P. T., Miles, M. J., Morris, V. J., O'Neil, M. A. & Sutherland, I. W. (1987). Effect of acetylation on the molecular interactions and gelling properties of a bacterial polysaccharide. *International Journal of Biological Macromolecules* **9**, 115–17.

Atkins, E. D. T., Niedusynski, I. A., Mackie, W., Parker, K. D. & Smolko, E. E. (1973). Structural components of alginic acid. II. Crystalline structure of poly-α-L-guluronic acid. Results of X-ray diffraction and polarized infra-red studies. *Biopolymers* **12**, 1879–87.

Dea, I. C. M., Morris, E. R., Rees. D. A., Welsh, E. J., Barnes, H. A. & Price, J. (1977). Associations of like and unlike polysaccharides: mechanism and specificity in galactomannans, interacting bacterial polysaccharides and related systems. *Carbohydrate Research* **57**, 249–72.

Dea, I. C. M., Clark, A. H. & McCleary, B. V. (1986). Effect of galactose substitution patterns on the interaction properties of galactomannans. *Carbohydrate Research* **147** 275–94.

McCleary, B. V., Clark, A. H., Dea, I. C. M. & Rees, D. A. (1985). The fine structures of carob and guar galactomannans. *Carbohydrate Research* **139**, 237–60.

Morris, E. R., Rees, D. A., Sanderson, G. R. & Thom, D. (1975). Conformation and circular dichroism of uronic acid residues in glycosides and polysaccharides. *Journal of the Chemical Society, Perkin Transactions* **2**, 1418–25.

Morris, E. R., Rees, D. A., Thom, D. & Boyd, J. (1978). Chiroptical and stoichiometric evidence of a specific primary dimerisation process in alginate gelation. *Carbohydrate Research* **66**, 145–54.

Morris, V. J., Franklin, D. & I' Anson, K. (1983). Rheology and microstructure of dispersions and solutions of the microbial polysaccharide from *Xanthomonas campestris* (xanthan gum). *Carbohydrate Research* **121**, 13–30.

Saito, H., Ohki, T. & Sasaki, T. (1977). A ^{13} C Nuclear magnetic resonance study of gel-forming $(1\rightarrow 3)$ β-D-glucans. Evidence of the presence of single helical conformation in a resilient gel of a curdlan type polysaccharide 13140 from *Alcaligenes faecalis* var. *myxogenes* IFO 13140. *Biochemistry* **16**, 908–14.

Shatwell, K. P. (1989). The influence of acetyl and pyruvic acid substituents on the solution and interaction properties of xanthan. Ph.D. thesis, Edinburgh University.

Shatwell, K. P., Sutherland, I. W. & Ross-Murphy, S. B. (1990). Influence of acetyl and pyruvate substituents on the solution properties of xanthan polysaccharide. In *Physical networks* (ed. W. Burchard & S. B. Ross-Murphy), pp.315–34. Elsevier, Amsterdam.

Chapter 9

Graber, M., Morin, A., Duchiron, F. & Monsan, P. F. (1988). Microbial polysaccharides containing 6-deoxysugars. *Enzyme and Microbial Technology* **10**, 198–205.

Sanderson, G. R. & Clarke, R. C. (1983). Gellan gum. *Food Technology* **37**, 63–70.

WHO (1987). *Toxicological evaluation of certain food additives and contaminants.* Cambridge University Press.

Chapter 10

Hughes, D. S., Teeuw, D., Cottrell, C. W. & Tollas, J. M. (1988). Appraisal of the use of polymer injection to suppress aquifer influx and improve volumetric sweep in a viscous (15 cP) oil reservoir. *Society for Petroleum Engineers Paper* 17400.

Kierulf, C. & Sutherland, I. W. (1988). Thermal stability of xanthan preparations. *Carbohydrate Polymers* **9**, 185–94.

Lund, T., Smidsrod, O., Stokke, B. J. & Elgsaeter, A. (1988). Controlled gelation of xanthan by trivalent chromic ions. *Carbohydrate Polymers* **8**, 245–56.

Nguyen, A.-L., Grothe, S. & Luong, J. H. T. (1988). Applications of pullulan in aqueous two-phase systems for enzyme production, purification and utilization. *Applied Microbiology and Biotechnology* **27**, 341–6.

Skjak-Braek, G. & Martinsen, A. (1989). Application of some algal polysaccharides in biotechnology. In *Seaweed resources in Europe* (ed. G. Blunden & M. Guiry). Heydon & Son, London.

Chapter 11

Allen, P. M., Fisher, A. D., Saunders, J. R. & Hart, C. A. (1987*a*). The role of capsular polysaccharide K21b of *Klebsiella* and of the structurally related colanic acid polysaccharide of *Escherichia coli* in resistance to phagocytosis and serum killing. *Journal of Medical Microbiology* **24**, 363–70.

Allen, P. M., Roberts, I. Boulnois, G. J., Saunders, J. R. & Hart, C. A. (1987*b*). Contribution of capsular polysaccharide and surface properties to virulence of *Escherichia coli* K1. *Infection and Immunity* **55**, 2662–8.

Lenoir, A. A., Pandey, J. P. & Granoff, D. M. (1988). Antibody responses of black children to *Haemophilus influenzae* type b polysaccharide–*Neisseria meningitidis* outer membrane protein conjugate vaccine in relation to the Km(1) allotype. *Journal of Infectious Diseases* **157**, 1242–5.

Index

Printed in the United States
By Bookmasters